General Editor
Jeff Daniels

COLLINS
London and Glasgow

First published 1986
Wm Collins Sons & Co Ltd

© Chevprime Ltd 1986

Produced by Chevprime Ltd
147 Cleveland St
London W1

ISBN 0 00 458836 3

Filmset by Computerised Typesetting Services Ltd

Printed by New Interlitho, Italy

Contents

Introduction
by Jeff Daniels — 4

List of abbreviations — 15

Racing Cars — 16

Index — 236

Introduction

Racing motor cars has always been a subject with its own fascination, almost from the time the motor car itself was invented. What better way could there have been for the early car manufacturers to demonstrate the superiority of their products than to show they were the fastest between two major European cities? That was how serious racing first began – and how it nearly ended, when the authorities stepped in to stop the Paris–Madrid race of 1903 when the competitors reached Bordeaux. It was an indication that cars had already developed to the stage where they were too fast to be driven competitively on ordinary public roads. The accidents of the Paris–Madrid were dangerous, to competitors and onlookers alike.

It was quickly realised that the alternative was to race on roads closed to the public – which, in effect meant some kind of circuit. The earliest circuit races were run in France, just as the big inter-city races had started mainly in Paris: this was because France was the cradle of technical development for the motor car before 1914. The first French Grand Prix was held at Le Mans in 1906, and it set the scene which has since become so familiar: the 'works' teams, the pits, and above all, car designs especially adapted for racing rather than road use. It was also, from the outset, the scene for intense national rivalry. The French quickly found their supremacy threatened by the Germans, in the shape of Mercedes, and the Italians with the mighty Fiats.

In those early years it was the popular belief that the way to make a car faster was to give it a bigger engine. Consequently, some of the pre-1914 racers had engines which were huge by any standards, of 10-litre capacity and more. Yet, as the engines became bigger so the cars became heavier and more difficult to drive. Peugeot was the first to understand that a smaller but more efficient engine could give superior power-to-weight ratio for better acceleration, while also enabling the designer to make the car lighter and faster round corners. That was the reasoning which led to the famous victories of Boillot in the French GPs of 1912 and 1913; and it is well worth looking at the 1912 Peugeot as possibly the first example of a 'real' racing car in the modern sense.

Needless to say, there was a reaction against the trend towards highly specialised designs for racing cars. Many people, especially in the years 1919–1939, thought the only valid form of racing involved the kind of car anyone could buy and run on the road. That feeling was the origin of the many sports car races which evolved in those years, especially the Le Mans 24 Hours and the Targa Florio. Eventually, and perhaps inevitably, all such events tripped over the obvious question: how do you define a sports car? Meanwhile, the most prestigious races -- the Grand Prix – continued to be disputed by cars whose sheer power and driving characteristics made them difficult if not impossible to drive anywhere but on a closed circuit. True, some of the best racing cars of the

1920s, like the Bugattis, were also sold in small numbers to road-going enthusiasts: but the 'works' racing cars were significantly more powerful and tricky to drive even so. The last vestige of road-going practice was swept away in 1925 with the abolition of the riding mechanics. These brave, if not foolhardy individuals, had served a real purpose in the earliest races (pumping pressure into the fuel tank and lubricant to the engine and the chassis, quite apart from dealing with the frequent punctures) but by the 1920s, they had become mere high-speed passengers.

There was another more significant argument which raged in the world of the 1920s racing car, and that was the question of 'formula'. It was easy enough for most people to accept that GP racing cars should not be constrained by the need to be 'street-legal' when they would never see a street in their running lives: the real argument lay in whether motor racing should be 'anything goes' or if the cars should be made to conform to some kind of specific formula. At first sight the idea of Formula Libre is simple and appealing: any kind of car is allowed as long as it has four wheels, and the first past the post wins. Yet there are equally strong objections to the idea. Such a free-for-all could lead to freakish designs. In a totally Formula Libre for instance there would be nothing to stop someone entering a vehicle so wide that it would be difficult or impossible to pass. More seriously, there is the risk that in a wide range of designs there would be only one or two that would strike the best balance, so that races would become mere processions. Far better to agree to a 'formula' to which all GP cars must conform.

Even in 1914 it was realised that the most practical formula combined a limit on engine capacity – to discourage repeated cycles of engine-growth of the kind seen up to 1912 – with a *minimum weight* (to discourage the building of anything frighteningly light and fragile). Through the 1920s such formulae applied and it was in those years that some of the classic battles were fought between cars like the Alfa Romeo P2 and the Bugatti Type 35. By the 1930s different formulae had been adopted. Formula 1, governing most Grands Prix, was in effect a Formula Libre from 1931 to 1937 and these years saw hectic technical development and rapidly escalating costs. The GP racing scene became one of national competition between the Italians and the Germans – Alfa Romeo and Maserati versus Auto-Union and Mercedes. The British and the French, whose governments were not inclined to back national prestige to the tune of millions of pounds, were completely squeezed out.

For all that, the 1930s had an excitement of their own, and it would be 40 years before Formula 1 cars again had the immense power of designs like the Auto-Union C-type and the Mercedes W125. It was also an era when the cult of the racing driver as a personality reached new heights, with the emergence of wizards like Nuvolari, Caracciola and Rosemeyer. The sight of these masters deploying

600bhp on skinny tyres, in days when the art of chassis design for good stability and handling was far from

1937 German Grand Prix at Nürburgring. Two Mercedes-Benz driven by Lang and Carcciola lead two Auto-Unions driven by Rosemayer and Muller.

fully understood, was a stirring one indeed. In the end, the technical supremacy of the Germans became almost depressing. Attempts to upset their success by adopting a new Formula 1 (taking advantage of the fact that the governing headquarters and political influence of motor sport remained, by virtue of early history, in Paris) were confounded; the Germans showed also that they were able to adapt to new regulations faster than anyone else.

After 1945 the racing scene had understandably lost much of its former interest. Such Grands Prix as there were tended to be won by the Italians, who had managed to preserve and develop some of their pre-war designs like the Alfa Romeo Tipo 158, and who soon supplemented them with new cars from Maserati and above all, Ferrari. The only opposition came from various British and French designs which were developed on the proverbial 'shoestring' budget. It was an era when, for lack of any significant number of competitive Grand Prix cars, many races were run to other formulae. The same situation had been seen before the war when British racing especially was full of one-off 'specials' and lightweight designs like the Alta and ERA. One result was that the number of 'junior' formulae quickly grew. Indeed, it was out of the 1950s Formula 3 (for cars with 500-cc engines) that the mid-engined layout emerged which was to change the shape of all racing cars from the 1960s onwards.

Eventually the Italian dominance of GP racing was challenged, first by cars like the Connaught and the Vanwall, and then by a new breed of British specialist, led by Cooper and Lotus. Jack Brabham's victories in the Cooper-Climax, together with those of Jim Clark and Graham Hill in the Lotus 25, ushered in a new kind of racing. A new stability of formula also emerged. In 1966 engines of 3-litre capacity (or 1.5-litres supercharged) were allowed, and this rule has remained essentially unchanged ever since. However, there have been many detailed rule changes brought about for a variety of reasons, including a concern for driver safety and the need to control the new exploitation of aerodynamics.

Perhaps the first real 'classic' of the 3-litre era was the Lotus 49, the first car to use the Ford-Cosworth V-8 engine which was destined to win more Grands Prix than any other power unit in history. Sadly, the 49 was also the last car driven to GP victory by Jim Clark before his untimely death. In the following years the GP car changed shape in several ways. Concern that 3-litre power would be too much for just two tyres to put on the ground resulted in a rash of four-wheel-drive cars like the Lotus 63; the trend was short-lived

1955 British Grand Prix. Stirling Moss (12), Fangio (10) both Mercedes. No 2 is the Maserati of Jean Behra.

The TAG-McLaren Turbo engine which is made by Porsche.

because designers discovered they could exploit aerodynamic download to make sure the rear wheels alone had enough traction.

To begin with, such downloads were applied simply by attaching 'wings' to the bodywork or even directly to the suspension, but after a series of accidents caused by wings falling off – notably a pair of

huge 'shunts' at Barcelona in the Spanish Grand Prix – severe restrictions were placed on such additions. Then Colin Chapman of Lotus discovered that even greater downforces could be achieved by subtle shaping of the underside of the car body, together with the use of 'skirts' to prevent air leakage. This was the principle behind the Lotus 72, another classic design, which led in turn to the current generation of GP winners like the McLarens and Brabhams.

The other major upheaval in GP racing came in the late 1970s when Renault exploited the provision within the formula for 1.5-litre supercharged engines. The challenge of the French company with its RS01 was so successful that today, all GP cars use turbocharged power units.

It should not be forgotten that while GP racing is the pinnacle as far as most people are concerned, there are many other kinds of racing car. In the USA, for example, formulae exist which have been drawn up to meet local traditions and conditions: the most famous American race beyond doubt is the Indianapolis 500. At the same time, sports car racing has continued to attract a lot of interest, though modern sports-racing cars bear little resemblance to anything that would be practical on the road. There was a time in the early 1970s when a generation of highly aerodynamic Ferrari and Porsche 5-litre sports cars proved, if anything, slightly faster even than the GP cars of the day; today however there is generally a 3-litre limit on engine capacity.

List of abbreviations

bhp	brake horse power
cc	cubic capacity
cm	centimetres
GP	Grand Prix
ifs	independent front suspension
ioe	inlet overhead exhaust
km/h	kilometres per hour
mph	miles per hour
mm	millimetres
ohc	overhead camshaft
ohv	overhead valves
rpm	revs per minute

CANNSTATT-DAIMLER Phoenix

Country of Origin: Germany
Date: 1899
Engine: four-cylinder, 5.5-litre water-cooled; side valves (automatic inlets); 24 or 28hp
Gears: four-speed
Capacity: 5,507cc
Bore & Stroke: 106 × 156mm
Maximum Speed: 72km/h (45mph)

This car had a short wheelbase and a high centre of gravity which made it one of the most dangerous racing cars of its time. Despite its speeding abilities (45mph was good going), it was only moderately successful in racing. Jellinek raced it at the Nice Speed Week in 1899 but apart from that it did not appear in any major races that year, although it won the Nice–Castelane hill climb in the touring car class. In the 1900 Nice–Marseilles race, Jellinek came in last place and a few days later, at the La Turbie hill climb, Wilhelm Bauer was killed when his car ran wide on a corner and smashed into some rocks. The car was never raced again.

PANHARD 70hp

Country of Origin: France
Date: 1902
Engine: four vertical cylinders; side valves (automatic inlets), three per cylinder
Gears: four-speed
Capacity: 13,672cc
Bore & Stroke: 160 × 170mm

Chassis: flitchplate frame, with engine bolted directly on to it

The Panhard 70hp won the big-car class in the 1902 Paris–Vienna race, driven by Henry Farman who also came overall second. In all no fewer than eight of these models took part, and de Knyff averaged 87.5km/h (54.5mph) for the first part of the race from Paris to Belfort. Another success was in the Circuit des Ardennes when Charles Jarrott won by defeating a 60-hp Mors.

PEUGEOT 7.6-litre Grand Prix

Country of Origin: France
Date: 1912
Engine: four vertical cylinders; four inclined ohvs, per cylinder operated by twin ohcs; 130bhp at 2,200rpm
Gears: four-speed
Capacity: 7,603cc
Bore & Stroke: 110 × 200mm

In the 1912 Grand Prix Georges Boillot drove this car to beat the chain-driven Fiats (which had nearly double the engine capacity) at 110.16km/h (68.45mph). Goux won the 1912 Coupe de la Sarthe, and the 1913 Indianapolis 500-Mile Race at 122.2km/h

(75.9mph). Peugeot was the first to introduce what became the standard features of motor racing engines – with the twin ohcs actuating four valves at the head of each cylinder, two inlets allowing the mixture of air and fuel to enter and two exhausts. During 1912 and 1913, this new feature led to Peugeot's dominance in GP racing, only challenged by Mercedes in 1914.

AUSTRO-DAIMLER Sascha
Country of Origin: Austria
Date: 1922

Engine: four-cylinder, short-stroke; twin ohcs; dual ignition; 45bhp at 4,500rpm
Gears: four-speed
Capacity: 1,100cc
Bore & Stroke: 68.3 × 75mm
Maximum Speed: 145km/h (90mph)
Chassis: front suspension by semi-elliptic springs. Rear suspension by cantilever springs
Dimensions: wheelbase 244cm (96in)
Brakes: four-wheel

The Sascha was named after its racing driver/film magnate sponsor, Count Alexander Kolowrat. The

car had an immediate success in the 1922 Targa Florio in the 1,100cc production class, when it took first and second places. In the racing class Alfred Neubauer came sixth. The legendary Saschas took 43 first and eight second places out of 52 starts in the 1922 season.

ALFA ROMEO P2

Country of Origin: Italy
Date: 1924
Engine: straight eight, supercharged, 2-litre; 134bhp at 5,200rpm; twin ohcs; single carburettor
Gears: four-speed
Capacity: 1,987cc
Bore & Stroke: 61 × 85mm
Maximum Speed: 217km/h(135mph)
Dimensions: wheelbase 259cm (102in)
The P2, of which six were built, was designed by Vittorio Jano who had been lured away from Fiat for the purpose. It was a landmark in GP racing, and was an immediate success, winning its first race at

Cremona and the 1924 French GP d'Europe at Lyons. In the same year having had Memini carburettors fitted, it won the Italian GP at Monza. By 1925 the P2s were producing 156bhp, they won the GP d'Europe at Spa, the Italian GP for the second year running and the Manufacturers' World Championship for Alfa Romeo. For the next five years the P2 went on winning a great number of racing events, and after a number of modifications was phased out in 1932.

BUGATTI Type 35

Country of Origin: France
Date: 1924
Engine: straight eight; single ohc; three valves per cylinder; twin Zenith carburettors; 95bhp at 5,000rpm
Gears: separate four-speed manual
Capacity: 1,991cc
Bore & Stroke: 60 × 88mm
Maximum Speed: 177km/h (110mph)
Chassis: pressed steel side members. Front suspension by semi-elliptic front springs. Rear suspension by reversed quarter-elliptic springs; friction dampers

Dimensions: wheelbase 240cm (94.5in). Track (front) 125cm (49in); track (rear) 119cm (47in)
Brakes: four-wheel drum integral with alloy wheels
This sensational model first appeared in the French GP of 1924, although its first season was less than successful. In 1925 it began to show its promise; it won the Rome GP, the Sicilian Targa Florio, which turned out to be the first of five successive wins and was the beginning of the Type 35's brilliant racing career.

SUNBEAM Grand Prix

Country of Origin: Great Britain
Date: 1924
Engine: six-cylinder; Roots-type supercharger, 2-litre; twin ohcs; 146bhp; Solex carburettor
Gears: four-speed
Capacity: 1,988cc
Bore & Stroke: 67 × 94mm
Maximum Speed: 209km/h (130mph)
Dimensions: wheelbase 259cm (102in)
The Sunbeam of 1924 was a development of the famous 1923 'Fiats in green paint' designed by Bertarione. The 1924 version had a supercharged

engine and a lowered chassis. Despite these modifications, and although it won the Spanish San Sebastian GP, this turned out to be the last GP victory of note by an all-British car and driver until 1955. A Sunbeam driven by Masetti came in third in the French GP in 1925.

STUTZ Black Hawk

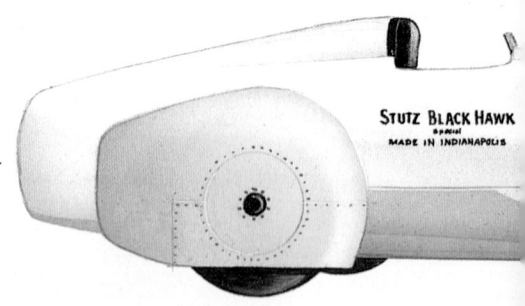

Country of Origin: USA
Date: 1927
Engine: straight eight; single ohc; twin carburettors; 95bhp at 3,200rpm
Gears: three-speed manual
Capacity: 4,888cc
Bore & Stroke: 82.5 × 114.3mm
Maximum Speed: 153km/h (95mph)
Chassis: pressed steel side members. Front and rear suspension comprising semi-elliptic leaf springs and friction dampers
Dimensions: wheelbase 333cm (131in). Track (front) 144cm (56.5in); track (rear) 149cm (58.5in)
Brakes: four-wheel drum

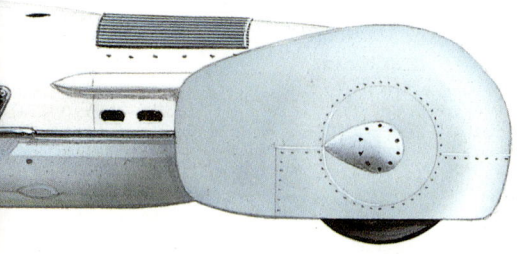

Designed by Moscovics and Lockhart, the Black Hawk had a very successful career in AAA races and in 1928 set a new stock-car record of 171km/h (106.5mph). A 3-litre Black Hawk 'Special' was designed and built in 1927 to attempt the Land Speed Record, its 3-litre capacity being the smallest swept volume of any vehicle built to attempt the record. This had two 1½-litre Miller engines with two superchargers and an estimated output of 385bhp at 7,500rpm. In 1928 Lockhart made two attempts at the record at Daytona, and on the second attempt reached 327.5km/h (203.45mph). Tragically, however, on the return, a tyre burst and Lockhart was killed in the crash.

ERA-DELAGE Grand Prix

Country of Origin: Great Britain/France
Date: 1927/51
Engine: ERA, six-cylinder; inclined oliv with push-rods; two-stage Roots supercharger
Gears: four-speed, Armstrong-Siddeley preselector
Capacity: 1,488cc
Bore & Stroke: 62.8 × 80mm
Maximum Speed: 225km/h (140mph)
Chassis: Delage channel-section frame. Independent front suspension by transverse leaf spring and upper wishbones. Rear suspension by rigid axle on semi-elliptic leaf springs.
Dimensions: wheelbase 250cm (98.5in)

This model was a successful combination of the 1927 Delage chassis and the engine of the first E-type ERA which crashed in the Isle of Man. The offset driving position and the external appearance of the car remained as before, but a new gearbox, brakes, shock absorbers, tyres and wheels were fitted. Raced by Tony Rolt it had a couple of successful seasons in Britain in 1951 and 1952, coming third at Goodwood behind a Tipo 159 Alfa Romeo and a $4\frac{1}{2}$-litre Ferrari.

INVICTA

Country of Origin: Great Britain
Date: 1931
Engine: Meadows six-cylinder; in-line ohv with push-rods; twin SU carburettors; dual ignition
Gears: four-speed
Capacity: 4,467cc
Bore & Stroke: 88.5 × 120.64mm
Maximum Speed: more than 150km/h (95mph)
Chassis: channel-section frame, underslung at rear. Front and rear suspension by semi-elliptic leaf springs, hydraulic dampers front and rear

Dimensions: wheelbase 300cm (118cm). Track 142cm (56in)
Brakes: four-wheel drum
A single-seater model was built for the BRDC 500-Mile Race in 1932. Built from components of this basic sports model it received a lot of publicity but was never put to the test because Roland Hebeler crashed it during a practice run and completely wrote it off.

DUESENBERG Wonder Bread Special

Country of Origin: USA
Date: 1933
Engine: Clemons, V-8; 4.5-litre; inclined ohvs with two ohcs; two carburettors
Gears: three-speed
Capacity: 4,376cc
Bore & Stroke: 88.4 × 89mm
Maximum Speed: 233km/h (145mph)
Chassis: channel-section frame. Front and rear suspension by semi-elliptic leaf springs
Dimensions: wheelbase 253cm (99.5in)

The Wonder Bread Special was built for Count Trossi, President of the Scuderia Ferrari, for use in European events. Its GP career was not very successful. Trossi lent it to Whitney Straight who drove it at Brooklands, but the chassis design meant that it was not an easy car to control, and many modifications were made to try to improve this defect. Its fastest speed at Brooklands was 222.6km/h (138.4mph) achieved by Whitney Straight.

MG K3 Magnette

Country of Origin: Great Britain
Date: 1933
Engine: MG, six-cylinder; ohvs with single ohc; supercharged; 140bhp
Gears: four-speed, ENV preselector
Capacity: 1,087cc
Bore & Stroke: 57 × 71mm
Maximum Speed: 201km/h (125mph)

Chassis: channel-section frame. Front and rear suspension by semi-elliptic leaf springs
Dimensions: wheelbase 239cm (94.25in)

The K3 Magnette was the last in a series of six-cylinder cars which began with the K1 Magnette saloon in 1932, followed almost immediately the K2 two-seater and then the supercharged K3. This remarkable car won its class in the 1933 Mille Miglia and won the RAC Tourist Trophy at Ards, driven by Tazio Nuvolari. They developed a reputation as sound and powerful racing cars. Today's surviving originals are collector's items.

NAPIER-RAILTON

Country of Origin: Great Britain
Date: 1933
Engine: Napier Lion aero engine; 12 cylinders in three banks of four; ohvs with ohc on each bank; 500bhp at 2,200rpm
Gears: three-speed
Capacity: 23,970cc
Bore & Stroke: 139.7 × 130.2mm
Maximum Speed: 274km/h (170mph)
Chassis: under-slung channel-section frame. Front suspension by semi-elliptic leaf springs. Rear suspension by double cantilever leaf springs
Dimensions: wheelbase 330cm (130in)

This car was designed particularly for record breaking and track racing. The chassis was designed by Reid Railton. It took many records at Montlhéry and Bonneville; it frequently broke the Brooklands lap record; it was Brooklands Champion in 1934, and won the Brooklands Track Drivers' Star for John Cobb in 1937. In 1935 it won the BRDC 500-Mile Race at 195km/h (121.3mph) and the BRDC's 500-Kilometre Race in 1937 at 204km/h (127mph).

ALFA ROMEO Tipo B

Country of Origin: Italy
Date: 1934
Engine: Alfa Romeo, eight cylinders, in line; inclined ohvs with two ohcs; twin Roots superchargers; 255bhp at 5,400rpm
Gears: three-speed
Capacity: 2,905cc
Bore & Stroke: 68 × 100mm
Maximum Speed: 233km/h (145mph)
Chassis: channel-section frame. Front suspension by semi-elliptic leaf springs. Rear suspension by semi-elliptic leaf springs with divided prop-shaft axle.
Dimensions: wheelbase 266cm (105in)
Brakes: drums, rod-operated

The Tipo B, also known as the 'monoposto', was the first great single-seater and remains one of the classic racing cars of all time. Its finest hour was in the 1935 German GP, when Italy's champion driver, the great Tazio Nuvolari, took on the might of the state-backed German team of Mercedes-Benz cars (in a slightly modified version to the one shown here) and scored an historic win.

APPLETON Special

Country of Origin: Great Britain
Date: 1934/39
Engine: Riley Nine; four cylinders; inclined ohvs with pushrods; Zoller supercharger; 118bhp
Gears: four-speed ENV preselector
Capacity: 1,089cc
Bore & Stroke: 60.3 × 95.2mm
Maximum Speed: 217km/h (135mph)
Chassis: channel-section frame. Front and rear suspension by semi-elliptic leaf springs
Dimensions: wheelbase 228cm (90in)

The Appleton Special began as a 1,100cc Maserati body with a tuned Riley Nine engine. Over the years the chassis was progressively shortened and the engine made more powerful. It reached its peak of performance in the 1938 season, when the Zoller supercharger was replaced by an Arnott and it developed 160bhp; the body was slimmer and the weight was reduced. The car's well-known feature was its ear-splitting exhaust note, which exceeded that of the ERAs of that time and double rear-wheels which considerably increased the car's roadholding.

BUGATTI Type 59

Country of Origin: France
Date: 1934
Engine: Bugatti, eight cylinders in line 3.3-litre; inclined ohv with two ohcs; Roots-type supercharger
Gears: four-speed
Capacity: 3,257cc
Bore & Stroke: 72 × 100mm
Maximum Speed: 257km/h (160mph)
Chassis: channel-section. Front suspension, semi-elliptic leaf springs. Rear suspension reversed quarter-elliptic leaf springs
Dimensions: wheelbase 269cm (106in)
This most elegant Bugatti model had several innovations including turning the crankshaft by bevel

gearing, a starting handle in the left-hand side of the chassis and a non-independent front end. In fact it marked the beginning of Bugatti's decline as a power in GP circles, and from 1935 to 1939 Bugattis were unsuccessful in GP racing. Bugatti eventually took them out of GP racing and into sports-car racing, where they were more successful.

MASERATI 8CM

Country of Origin: Italy
Date: 1934
Engine: Maserati, eight cylinders in line; two valves per cylinder; inclined ohv with two ohcs; supercharged; 210bhp at 5,500rpm
Gears: four-speed
Capacity: 2,992cc
Bore & Stroke: 69 × 100mm
Maximum Speed: 233km/h (145mph)
Chassis: channel-section. Front and rear suspension, semi-elliptic leaf springs
Dimensions: wheelbase 257cm (102.75in)
Brakes: hydraulic drum

The 8CM was introduced in the 1933 season when the incredibly narrow-bodied version (only 62cm/24.5in wide) had a series of brilliant victories. However, it was a difficult car to manage and Campari and Borzzachini were both killed in the Monza GP. The legendary Nuvolari continued to race the narrow-bodied car but wrecked it in the Bordino Cup race at Alessandria the following year. He continued racing in the wider-bodied version coming second in the Coppa Acerbo at Pescara. The 8CM was then replaced by the Tipo 34.

FRAZER-NASH Single seater

Country of Origin: Great Britain
Date: 1935
Engine: Gough; four cylinders in line; ohv with one ohc; two valves per cylinder; two Centric superchargers; 160bhp at 7,000rpm
Gears: four-speed, chains
Capacity: 1,496cc
Bore & Stroke: 69 × 100mm
Maximum Speed: 201km/h (125mph)
Chassis: channel-section frame. Front suspension by straight beam on cantilevered half-elliptic springs. Rear suspension by quarter-elliptic springs

Dimensions: wheelbase 274cm (108in)
Brakes: cable-operated drum

Only three of these cars were ever built, but one made history at Shelsley Walsh in 1937. Driven by A. F. P. Fane it did the hill-climb in 38.77 seconds. These cars were the epitome of the spartan structure and construction of British post-vintage performance cars.

SQUIRE Single Seater

Country of Origin: Great Britain
Date: 1935
Engine: Anzani R1; four-cylinder; 1½-litre; twin ohc; Roots supercharger; inclined ohvs
Gears: four-speed; Wilson preselector gearbox
Capacity: 1,496cc
Bore & Stroke: 69 × 100mm
Maximum Speed: 201km/h (125mph)
Chassis: channel-section. Front and rear suspension, semi-elliptic leaf springs, damped by thermostatically controlled shock absorbers
Dimensions: wheelbase 259cm (102in)
Brakes: hydraulic drum

The Squire was one of the fastest sports cars of its time with its extremely powerful engine, and the single-seater version was built on the sports car chassis. Luis Fontes raced it at Brooklands in 1935, but it was outpaced by Altas and ERAs. It retired from the British Empire Trophy when the engine blew up and from the 500-Mile Race when the chassis broke. It only ran once more after that, at Brooklands when it gained third place, then was converted back into a two-seater road car.

TROSSI MONACO

Country of Origin: Italy
Date: 1935
Engine: Monaco-Aymini, eight radial cylinders, double piston, two-stroke; Zoller Type M160 superchargers
Gears: four-speed
Capacity: 3,982cc
Bore & Stroke: 65 × 75mm
Maximum Speed: 241km/h (150mph)
Chassis: tubular spaceframe. Independent front suspension with lever-operated horizontal coil. Independent rear suspension with lever-operated horizontal coil
Dimensions: wheelbase 228cm (90in)

This car was of a very unusual design for its time. Constructed at Count Trossi's castle workshops at Biella in Northern Italy, the chassis frame was made of small-diameter tubing constructed along aircraft lines, a type of design which is now very common. The engine, mounted on the nose, operated on the split-single principle, air-cooled, and the front wheels were driven by the superchargers. It was a new concept in racing but, because the pace of development in GP racing was so rapid at this time, this one off special had little chance to succeed, and was never raced.

VALE Special

Country of Origin: Great Britain
Date: 1935
Engine: Coventry-Climax, 1½-litre; ioe valve arrangement; Centric supercharger developing 97bhp at 5,700rpm
Gears: four-speed
Capacity: 1,496cc
Bore & Stroke: 69 × 100mm
Maximum Speed: 201km/h (125mph)
Chassis: channel-section. Front and rear suspension comprising semi-elliptic leaf springs, underslung

Dimensions: wheelbase 213cm (84in)
Steering: separate drag links to each front wheel, no track rod
This car was successfully raced by Ian Connell and others who collected almost every available award in sporting events in 1934–36. Unfortunately, the Vale company's financial position prevented it from actively supporting these essays into the racing arena.

AUSTIN Twin-cam

Country of Origin: Great Britain
Date: 1936
Engine: Austin, four-cylinder; inclined ohvs with two ohcs; Roots supercharger; 116bhp at 8,500rpm
Gears: four-speed
Capacity: 744cc
Bore & Stroke: 60.3 × 65.1mm
Maximum Speed: 201km/h (125mph)
Chassis: channel-section frame. Front suspension by transverse leaf spring. Rear suspension by splayed quarter-elliptic leaf springs
Dimensions: wheelbase 208cm (82in)

Although, when it first appeared in 1936, there were many problems with fuel starvation and burned pistons, this neat and beautiful little car had a fairly successful racing career up until the start of World War II. A team of three were built and were driven by Driscoll, Goodacre, Dodson and Hadley. They raced mainly in Britain, and were particularly successful at Donington Park and Crystal Palace. In 1937 they won the Junior Trophy (Donington) and four races on Coronation Day; while at Crystal Palace they won the Crystal Palace Trophy and the Imperial Trophy. Other triumphs were at Shelsley Walsh, Craigantlet and at Brooklands where they won the Relay Race at a record 170km/h (105.6mph) in 1937.

AUTO UNION C-type

Country of Origin: Germany
Date: 1936
Engine: Auto Union, V-16; inclined ohv with single ohc and horizontal pushrods; Roots supercharger; 520bhp
Gears: four-speed
Capacity: 6,010cc
Bore & Stroke: 75 × 85mm
Maximum Speed: 314km/h (195mph)

Chassis: tubular ladder-type frame. Independent front suspension by trailing arms and transverse torsion bars. Independent rear suspension by swing axles and longitudinal torsion bars
Dimensions: wheelbase 290cm (114.5in)
Rosemayer had a triumphant 1936 season with these cars, winning the Eifel, German, Pescara, Swiss and Italian Grand Prix, and the European Championship. Varzi also won the Tripoli Grand Prix. In 1937 Rosemeyer again had a string of victories, winning the Eifel, Pescara and Donington GPs and the Vanderbilt Cup in the USA. Hasse won the 1937 Belgian GP, and von Delius won the Grosvenor GP in South Africa.

ERA B-type

Country of Origin: Great Britain
Date: 1936
Engine: ERA, six-cylinder; inclined ohv with push-rods; Jamieson-Roots supercharger
Gears: four-speed, ENV preselector
Capacity: 1,488cc
Bore & Stroke: 57.5 × 95.2mm
Maximum Speed: 209km/h (130mph)
Chassis: channel-section. Front and rear suspension by semi-elliptic leaf springs
Dimensions: wheelbase 239cm (94in)

The B-type ERAs gained a reputation in voiturette racing between 1935–7. One of the most successful was that driven by Prince Bira of Thailand which won the International Trophy in 1936, came second in the RAC 1,500cc Isle of Man Race, winning at Monaco and coming third in the Eifelrennen. Other B-types also scored notable successes.

ALTA Voiturette

Country of Origin: Great Britain
Date: 1937
Engine: Alta, 1½-litre, four-cylinder; inclined ohv with twin ohcs; supercharged
Gears: four-speed ENV preselector
Capacity: 1,488cc
Bore & Stroke: 68.75 × 100mm
Maximum Speed: 209km/h (130mph)
Chassis: channel-section frame. Independent front and rear suspension by vertical coil springs and sliding pillars
Dimensions: wheelbase 244cm (96in)

The Alta, master-minded by Geoffrey Taylor, was one of the first British racing cars to do away with the conventional radiator shell and substitute it with a radiator block behind a curved-nose cowling. The Alta engines were built down to a realistic price and so left something to be desired in terms of safety and reliability over long periods. The 1937 version, of which three were built, was raced successfully by George Abecassis despite these drawbacks.

MERCEDES-BENZ W125 Grand Prix

Country of Origin: Germany
Date: 1937
Engine: eight cylinders, in line; four ohvs per cylinder, operated by twin ohcs with finger rockers; single triple-choke Mercedes-Benz carburettor; Roots supercharger; 600bhp at 5,800rpm
Gears: four-speed, all-indirect, manual gearbox; no synchromesh
Capacity: 5,660cc
Bore & Stroke: 94 × 102mm
Maximum Speed: 338km/h (210mph)
Chassis: separate tubular steel frame, with oval tube side members and tubular bracings. Independent

front suspension by coil springs and wishbones. De Dion rear suspension by torsion bars with radius arms and sliding block of de Dion tube. Lever arm hydraulic dampers
Dimensions: wheelbase 279cm (110in)
Steering: worm-and-nut
Brakes: four-wheel, hydraulically operated drum, all outboard
This magnificent and remarkable car was used for only one season winning seven out of 13 events entered, and in leading positions at the finish of the other six.

ERA E-type

Country of Origin: Great Britain
Date: 1939
Engine: ERA, six-cylinder; inclined ohv with pushrods; Zoller supercharger; 260bhp
Gears: four-speed, synchromesh
Capacity: 1,488cc
Bore & Stroke: 62.8 × 80mm
Maximum Speed: 241km/h (150mph)
Chassis: tubular. Independent front suspension by Porsche trailing arms and torsion bars. Rear suspension, de Dion with torsion bars
Dimensions: wheelbase 262cm (103in)
Brakes: hydraulic

The E-type's career started full of promise in 1939 but degenerated into farce fairly rapidly. It was lighter than the A,B,C and D-type ERAs, with new synchromesh gearbox and de Dion rear axle. It ran only once in 1939, at Albi, where it crashed after running very well. In subsequent races it either failed to start at all or had to retire. In later years this record of failure continued. Whitehead drove it in Turin in 1946 when it retired with supercharger trouble. In races in 1947 and 1948 it had to retire from virtually every race it entered. It had better success in 1949, with two second places, two thirds and a fifth.

RAILTON-MOBIL

County of origin: Great Britain
Date: 1938
Engine: two Napier-Lion, 24-litre, 12-cylinder engines; ohvs with a single ohc to each bank
Gears: three-speed
Capacity: 23,970cc
Bore & Stroke: 139.7 × 130.3mm
Maximum Speed: 644 km/h (400mph)
Chassis: box-section backbone. Independent front suspension, rigid axle on coil springs
Dimensions: wheelbase 411.5cm (162in)

The Railton-Mobil was very different to other record-breaking cars and is included here because of the breakthrough in the use of four-wheel drive, independent front suspension and a backbone chassis frame. It had one of the most perfect aerodynamic bodies ever built, and was detachable in one piece for servicing. Driver John Cobb achieved his first record in 1938 at a speed of 563.6km/h (350.2mph). He improved on this in 1947 with a mean speed of 594.9km/(394.2mph)

MULTI-UNION II

Country of Origin: Great Britain/Italy
Date: 1939
Engine: Alfa Romeo, eight cylinders, in line; inclined ohv with two ohcs; two superchargers; 340bhp at 6,500rpm

Gears: four-speed
Capacity: 2,947cc
Bore & Stroke: 68.5 × 100mm
Maximum Speed: 257km/h (160mph)

Chassis: channel-section frame. Independent front suspension by Tecnauto trailing arm and coil spring. Rear suspension, rigid axle on coil springs
Dimensions: wheelbase 270cm (106.5in)
Brakes: Lockheed hydraulic
The Multi-Union won the Brooklands Outer Circuit Handicaps in 1928 at 204.4km/h (127mph) and 214km/h (133mph). Its suspension was modified using Tecnauto ifs, and coil springs replaced leaves at the rear. This improved 1939 model was set to beat the Napier-Railton's Outer Circuit Record at 230.76km/h (143.44mph), but a hole in a piston crushed this hope.

ALFA ROMEO Tipo 158 'Alfetta'

Country of Origin: Italy
Date: 1947/51
Engine: Alfa Romeo, eight cylinders in line; inclined ohv with two ohcs; two-stage Roots supercharging; 170bhp (later developed to over 380 bhp)
Gears: four-speed
Capacity: 1,479cc
Bore & Stroke: 58 × 70mm
Maximum Speed: 290km/h (180mph)

Chassis: tubular. Independent front suspension by trailing arms and transverse leaf spring. Independent rear suspension by swing axles and transverse leaf spring
Dimensions: wheelbase 250cm (98.5in)

It had a highly successful season in 1950 driven by Alfa Romeo's famous 'Three Fs' team of Fagioli, Farina and Fangio, winning ten Grand Prix.

TALBOT-LAGO
Type 26C

Country of Origin: France
Date: 1948
Engine: Talbot-Lago, six-cylinder; inclined ohv operated by pushrods and rockers; three carburettors; two ohcs; 240bhp at 4,700rpm
Gears: four-speed, Wilson preselector
Capacity: 4,485cc
Bore & Stroke: 93 × 110mm

Maximum Speed: 265km/h (165mph)
Chassis: box-section frame. Independent front suspension by wishbones and transverse leaf spring. Rear suspension, rigid axle on semi-elliptic leaf springs
Dimensions: wheelbase 250cm (98.5in)
This car's debut was in the 1948 Monaco GP. In 1949, driven by Rosier, it won the Belgian GP. Chiron won the French GP in the same year, when the 26C also had several other substantial victories. Rosier won the Dutch GP in 1950.

ALLARD Sprint

Country of Origin: Great Britain
Date: 1949
Engine: Steyr V-8; eight Amal carburettors; inclined ohv with pushrods
Gears: three-speed
Capacity: 3,600cc
Bore & Stroke: 79 × 92mm
Maximum Speed: 225km/h (140mph)

Chassis: channel-section. Independent front suspension with transverse leaf spring and swing axles. Rear suspension, de Dion with coil springs
Dimensions: wheelbase 254cm (100in)
Sidney Allard was a very skilful builder and driver of racing cars. He built his first Allard car in the 1930s, using a Ford engine and chassis, and Bugatti body and steering. The Sprint won the RAC Hill-Climb Championship in 1949.

HWM Formula 2

Country of Origin: Great Britain
Date: 1950
Engine: Alta, four-cylinder; inclined ohv with two ohcs; two double-choke Weber carburettors
Gears: four-speed ENV preselector
Capacity: 1,970cc
Bore & Stroke: 83.5 × 90mm
Maximum Speed: 209km/h (130mph)
Chassis: tubular frame. Independent front and rear suspension by transverse leaf spring and wishbones

Dimensions: wheelbase 234cm (92in)
Hersham & Walton Motors' cars were very successful in European racing in the 1950s. They used the Alta engine, and the HWM team had a very successful season in 1950.

MASERATI-MILAN Formula 1

Country of Origin: Italy
Date: 1950
Engine: Maserati-Milan, four-cylinder, 1½-litre; inclined ohvs with two ohcs; two-stage supercharging; 305bhp
Gears: four-speed
Capacity: 1,490cc
Bore & Stroke: 78 × 78mm
Maximum Speed: 257km/h (160mph)

Chassis: oval tube frame. Independent front suspension by double wishbones and torsion bars. Independent rear suspension by trailing arms and transverse leaf spring

Dimensions: wheelbase 251cm (98.75in)
These cars, produced by the Milan Scuderia, were in fact modified Maserati 4CLT/48s. The first version, which appeared in 1949, was fast but not reliable, and further development work was done. The brand new 1950 model, with its Speluzzi-built engine, raced in the 1950 season but again without success. The project was by this time becoming both complicated and expensive and the Scuderia Milan gradually let it die.

FERRARI 375

Country of Origin: Italy
Date: 1951
Engine: Ferrari; V-12; inclined ohv with one camshaft to each bank; three downdraught four-choke carburettors; 300bhp at 7,000rpm
Gears: four-speed
Capacity: 4,498cc
Bore & Stroke: 80 × 74.5mm
Maximum Speed: 290km/h (180mph)

Chassis: tubular. Independent front suspension by double wishbones and transverse leaf spring. Rear suspension, de Dion with transverse leaf spring
Dimensions: wheelbase 228cm (90in)
In 1951 this car ended the unbroken reign of the Alfa Romeo 158/159 when Froilan Gonzalez won the British GP at Silverstone. In the same year Ascari won the German and Italian GPs.

JAGUAR C-type

Country of Origin: Great Britain
Date: 1951
Engine: straight six; twin ohc; twin SU carburettors; 210bhp at 5,800rpm
Gears: four-speed manual
Capacity: 3,442cc
Bore & Stroke: 83 × 106mm
Chassis: spaceframe. Independent front suspension comprising wishbones, torsion bars and dampers. Rear suspension by rigid axle, radius arms transverse torsion bar, dampers

Dimensions: wheelbase 244cm (96in). Track 129cm (51in)
Brakes: four wheel drum
The C-type was developed by Jaguar for racing at Le Mans. It had a superb aerodynamic body designed by Malcolm Sayer, and new tubular chassis frame. C-types won the Le Mans 24-Hour Race and came first, second and fourth in the Tourist Trophy in 1951.

FERRARI 500

Country of Origin: Italy
Date: 1952
Engine: Ferrari, four-cylinder; inclined ohv with two ohcs; horizontal carburettors; 170bhp
Gears: four-speed
Capacity: 1,985cc

Bore & Stroke: 90 × 78mm
Maximum Speed: 241km/h (150mph)
Chassis: tubular. Independent front suspension by double wishbones and transverse leaf spring. Rear suspension, de Dion with transverse leaf spring
Dimensions: wheelbase 220cm (86.5in)

Developed for Formula 2 racing, the Ferrari 500 was virtually unbeatable in the 1952 and 1953 seasons. Alberto Ascari took two successive Drivers' World Championship titles; in 1952 he won six GPs and in 1953 he won five out of eight events. Also in 1953 Mike Hawthorn won the French GP and Farina the German event for the Prancing Horse team.

JAGUAR D-type

Country of Origin: Great Britain
Date: 1953
Engine: straight six; twin ohc; three Weber twin-choke carburettors; 245bhp at 5,500 rpm
Gears: four-speed manual
Capacity: 3,442cc
Bore & Stroke: 83 × 106mm
Chassis: central section magnesium-alloy monocoque. Tubular frames front and rear. Independent front suspension comprising wishbones, torsion bars, dampers, anti-roll bar. Rear suspension comprising rigid axle, trailing arms, transverse torsion

bar, dampers
Dimensions: wheelbase 230cm (90.5in). Track (front) 127cm (50in); track (rear) 122cm (48in)
Brakes: four-wheel disc

The D-type was created to challenge the racing might of Ferrari, Maserati, Lancia and Mercedes-Benz. The chassis was built on aviation principles. The panels were fixed to internal bulkheads to form the main load-bearing members of the frame, without having to use separate girder members. The D-type was aimed primarily at Le Mans and came second in 1954; it won the 24-Hour Race in 1955, 1956 and 1957. D-types also won the Sebring 12-Hour Race in Florida, USA in 1955, and Watkins Glen in 1955, 1956 and 1957. About 50 competition D-types were built.

CONNAUGHT 'B' Type

Country of Origin: Great Britain
Date: 1954
Engine: Alta, four-cylinder; inclined ohv with two ohcs; two double-choke Weber carburettors
Gears: four-speed ENV preselector
Capacity: 2,470cc
Bore & Stroke: 93.5 × 90mm
Maximum Speed: 249km/h (155mph)
Chassis: tubular frame. Independent front suspension by double wishbones and coil springs. Rear suspension, de Dion with torsion bars
Dimensions: wheelbase 229cm (90in)
This car made history in 1955 by being the first all-British car/driver combination to win a GP race since 1924; Tony Brooks, an amateur driver, took on the

Maserati works team at Syracuse and defeated them. The B Type was known as the 'Syracuse' model from then on. Initially the bodywork was an all-enveloping streamlined type, but this was later replaced by a more conventional slipper body, and it was this form which won at Syracuse. Unfortunately the Connaught company never raised sufficient financial support, and despite the creative technical and design ideas which the model B inspired, the company collapsed in 1957 and the Connaught's career was finished. Some survivors still take part in historic racing.

LANCIA D50

Country of Origin: Italy
Date: 1954
Engine: Lancia, V-8; inclined ohv with two ohcs per bank; for downdraught double-choke Weber carburettors; 295bhp at 8,500 rpm
Gears: five-speed
Capacity: 2,489cc
Bore & Stroke: 73.6 × 73.1mm
Maximum Speed: 274km/h (170mph)
Chassis: tubular spaceframe, with engine a stressed member. Front suspension comprising double wishbones and coil springs. Rear suspension, de Dion with transverse leaf springs

Dimensions: wheelbase 228cm (90in)

The D50 was a very small and light car to drive and the driving team of Ascari, Villoresi and Castellotti proved it to be extremely fast. It had very good handling but past a certain point suddenly tended to lose adhesion and the car would go into a spin. Despite this problem, Ascari won at Turin in 1954 and Pau in 1955. Ascari was killed subsequently in a Ferrari at Monza and the team never recovered. Lancia ran into enormous financial difficulties and the racing section was handed over to Ferrari late in 1955.

MASERATI 250F

Country of Origin: Italy
Date: 1954/7
Engine: Maserati, six cylinders; inclined ohv with two ohcs; three double-choke Weber carburettors; 260bhp
Gears: four- or five-speed
Capacity: 2,493cc
Bore & Stroke: 84 × 75mm
Maximum Speed: 274km/h (170mph)
Chassis: tubular spaceframe. Independent front suspension by double wishbones and coil springs. Rear suspension, de Dion with transverse leaf spring
Dimensions: wheelbase 228cm (90in)
Brakes: hydraulic drum

The 250F won the first race for which it was entered, the Argentine Grand Prix in 1954; it was driven by Fangio, who also drove a 250F to victory in the Belgian Grand Prix in the same year. In 1956 Stirling Moss won the Monaco Grand Prix. Then in 1957 Fangio took the Championship with the 250F by winning the Argentine, Monaco, French and German Grand Prix. This last race clinched Fangio's fifth and final World Drivers' Championship.

MERCEDES-BENZ W196 Grand Prix

Country of Origin: Germany
Date: 1954
Engine: Daimler-Benz M196, eight cylinders, in line; inclined ohvs desmodromic, two ohcs; Bosch fuel injection; 257bhp at 8,250rpm
Gears: five-speed, synchromesh, manual
Capacity: 2,496cc
Bore & Stroke: 76 × 68.8mm
Maximum Speed: 282km/h (175mph)

Chassis: tubular spaceframe. Independent front suspension by wishbones and torsion bars. Independent rear suspension by low-point swing axles and torsion bars
Dimensions: wheelbase 235cm (92.5in)
Steering: worm-and-sector
Brakes: hydraulic drum; inboard at both front and rear

The most extraordinary feature of this extraordinary model was the desmodromic valve gear, in which the valves were positively opened *and* closed by cams. By the end of 1955 the engine could produce 290bhp at 8,500rpm. It was an extremely reliable machine. The sports car body was intended for high-speed circuits such as Rheims and Monza

BRM P25

Country of Origin: Great Britain
Date: 1955
Engine: four cylinders in line; two valves per cylinder; two ohcs; 272bhp at 8,500rpm (on AvGas)
Capacity: 2,491cc
Bore & Stroke: 102.9 × 74.9mm
Chassis: tubular spaceframe. Independent front suspension by wishbones and coil springs. Rear suspension de Dion with coil spring struts
Brakes: disc

Designed by Stuart Tresilian, the P25 ran between 1955 and 1959. It was incredibly fast but handling was a problem and there were rear brake troubles and persistent valve failures. It was modified several times to try to solve these problems. In 1959 Behra won the non-Championship Caen GP and The Silverstone Trophy race, and in 1959 Bonnier scored BRM's first ever Championship victory in the Dutch GP.

GORDINI Formula 1

Country of Origin: France
Date: 1955
Engine: Gordini, eight cylinders; inclined ohv with two ohcs and rockers; four double-choke Weber carburettors; 230bhp
Gears: five-speed
Capacity: 2,475cc
Bore & Stroke: 75 × 70mm
Maximum Speed: 249km/h (155mph)
Chassis: tubular frame. Independent front and rear suspension by Watt-link mechanism and torsion bars

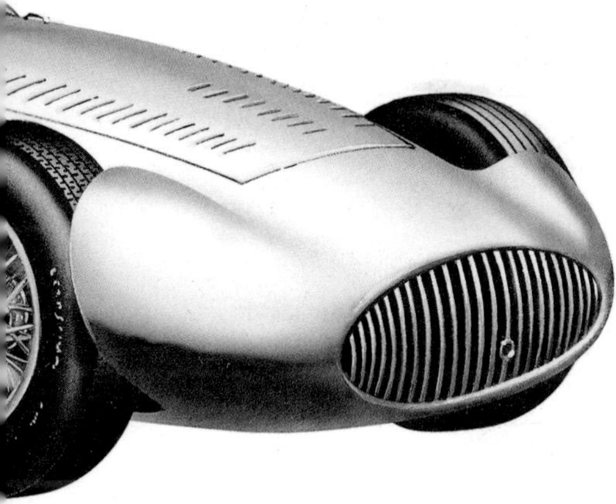

Dimensions: wheelbase 225cm (88.6in)
Brakes: Messier disc (rear inboard)
This car which appeared in late 1955, was a breakaway from previous Gordini six-cylinder models. It was larger and heavier, and the engine was not powerful enough to compete successfully with other makes. It staggered through two seasons and then quietly disappeared from the GP scene.

VANWALL Formula 1

Country of Origin: Great Britain
Date: 1958
Engine: Vandervell, four-cylinder; inclined ohv with two ohcs; Bosch fuel injection; 285 bhp
Gears: five-speed
Capacity: 2,490cc
Bore & Stroke: 96 × 86mm
Maximum Speed: 282km/h (175mph)

Chassis: tubular spaceframe. Independent front suspension by wishbones and coil springs. Rear suspension, de Dion with coil springs
Dimensions: wheelbase 229cm (90in)
Brakes: disc

Driven by Stirling Moss, Vanwalls won both the Pescara and Italian GPs and finally won the Manufacturer's Championship in 1958. The 1958 team of Moss, Brooks and Lewis-Evans was one of the finest GP racing teams ever. They won six of nine World Championship races and defeated both Ferrari and Maserati.

ASTON MARTIN F1 DBR4/250

Country of Origin: Great Britain
Date: 1959
Engine: Aston Martin, six-cylinder; inclined ohvs with two ohcs; 280bhp at 8,250rpm
Gears: five-speed
Capacity: 2,492cc
Bore & Stroke: 83 × 76.8mm
Maximum Speed: 257km/h (160mph)
Chassis: tubular spaceframe. Independent front suspension by double wishbones and coil springs. Rear suspension, de Dion with torsion bars
Dimensions: wheelbase 228cm (90in)

In 1959, Salvadori gave this car a promising start by coming second in the International Trophy at Silverstone, at an average speed of 164.8km/h (102.4mph). But subsequent performance was not encouraging, coming sixth in the British GP, 10th in the Italian GP and sixth and eight in the Portuguese GP. The cars were modified in the following year to conform to the 3-litre Tasman Formula and were more successful in Australian racing.

STANGUELLINI Formula Junior

Country of Origin: Italy
Date: 1959
Engine: Fiat, four-cylinder; pushrod ohv; two double-choke Weber carburettors
Gears: four-speed
Capacity: 1,089cc
Bore & Stroke: 68 × 75mm
Maximum Speed: 177km/h (110mph)
Chassis: tubular. Independent front suspension, upper transverse leaf spring and lower wishbones. Rear suspension, rigid axle on coil springs.

Formula Junior was introduced in Italy in 1958 and Stanguellini built a large batch of front-engined cars which were an immediate success. In 1959, when formula Junior went international Michel May came first at Monaco, Solitude and the Nurburgring, Bordeu won at Monza, von Trips won the Eifelrennan and Bandini won at Innsbruck. Despite this enormous success the Stanguellini was eclipsed in 1960 by the rear-engined Coopers and Lotuses.

COOPER-CLIMAX T53 GP

Country of Origin: Great Britain
Date: 1960
Engine: Coventry-Climax, four-cylinder 2½-litre; inclined ohv with two ohcs; two double-choke Weber carburettors
Gears: five-speed
Capacity: 2,495cc
Bore & Stroke: 94 × 89.9mm
Maximum Speed: 282km/h (175mph)
Chassis: tubular spaceframe. Independent front suspension by wishbones and coil springs. Independent rear suspension by transverse leaf spring and wishbones
Dimensions: wheelbase 231cm (91in)

The T53 won the Manufacturers' Championship in 1959 and 1960, and from then on it was obvious that the future was with rear-engine layout. The Cooper-Climax was being developed as early as 1957; increasingly larger engines were used, making the 1960 version much stronger, heavier and more complicated than the original model.

ELVA Formula Junior

Country of Origin: Great Britain
Date: 1960
Engine: DKW, three-cylinder; two-stroke porting valves
Gears: four-speed
Capacity: 986cc
Bore & Stroke: 74 × 76mm
Maximum Speed: 177km/h (110mph)
Chassis: tubular spaceframe. Independent front suspension by wishbones and coil springs. Independent rear suspension by wishbones, radius arms and coil springs
Dimensions: wheelbase 226cm (89in)

Frank Nicols' popular Elva (from the French *elle va* meaning 'she goes') was one of the first British cars to take part in Formula Junior racing when it was introduced in 1960. The DKW-engined Elva had some success at first, but was quickly overtaken by the rear-engined Lotus and Lola machines. This was partly because the Elva's engine was unreliable when highly stressed. Nicols tried to remedy this by putting the DKW engine behind the driver, and later inserted BMC and Ford engines in the search for greater reliability.

SCARAB Formula 1

Country of Origin: USA
Date: 1960
Engine: Scarab, four-cylinder; inclined ohvs, with two ohcs and desmodromic operation; Hilborn fuel injection
Gears: four-speed
Capacity: 2,441cc
Bore & Stroke: 95.25 × 85.73mm
Maximum Speed: 249km/h (155mph)
Chassis: tubular spaceframe. Independent front and rear suspension by double wishbones and coil springs
Dimensions: wheelbase 229cm (90in)

Lance Reventlow, the builder and designer of the Scarab F1 car, had previously built sports cars which had had several successes in American amateur racing. His design for the 1959 F1 car derived from contemporary European designs, with the chassis frame made of small-diameter tubing, and the engine having desmodromic valve gear. The car raced for the first time in 1960 but was not successful because new design developments had already occurred, including the rear-engined revolution. Other attempts by Reventlow to build racing cars also failed to keep up with the swift pace of developments, and within two years Reventlow had closed down his factory and left the racing scene.

FERGUSON P99 Formula 1

Country of Origin: Great Britain
Date: 1961
Engine: Coventry-Climax, four-cylinder; inclined ohv with two ohcs; two double-choke Weber carburettors
Gears: five-speed, Colotti
Capacity: 1,498cc
Bore & Stroke: 81.9 × 71.1mm
Maximum Speed: 241km/h (150mph)
Chassis: tubular spaceframe. Independent front and rear suspension by wishbones and coil springs; driveshafts
Dimensions: wheelbase 228cm (90in)
Brakes: disc, with anti-lock hydraulics

The P99 four-wheel drive model was built at the Ferguson Research Works near Coventry in order to test the Dunlop Maxaret anti-lock brake system, which was used for many years on aircraft. The engine was front mounted at a time when rear-engined layouts were increasingly used. With four-wheel drive the weight of the vehicle has to be evenly distributed over the four wheels to achieve the best results. The handling proved to be well up to expectations, particularly in the wet and, driven by Stirling Moss, it won the Oulton Park Gold Cup race in September 1961.

BRABHAM BT3 Formula 1

Country of Origin: Great Britain
Date: 1962
Engine: Coventry-Climax, V-8; inclined ohv with two ohcs; two valves per cylinder; 187 bhp at 8,500rpm
Gears: six-speed Colotti
Capacity: 1,494cc
Bore & Stroke: 63 × 60mm
Maximum Speed: 257km/h (160mph)

Chassis: tubular spaceframe. Front suspension by double wishbones and coil springs. Rear suspension by wishbone, link, radius rods and coil springs
Dimensions: wheelbase 231cm (91in)
Brakes: disc
This car made its debut in the 1962 German GP, but was forced to retire having run only two-thirds of the race. 1963 was more successful, with Jack Brabham driving to victory in the Australian, Austrian and Solitude GPs. The successes continued through the 1964 season with Brabham winning the *Daily Express* Trophy, Siffert winning the Mediterranean GP and Gurney the Mexican GP.

BRM TYPE 56 Formula 1

Country of Origin: Great Britain
Date: 1962
Engine: BRM, V-8; inclined ohv with four ohcs; Lucas fuel injection
Gears: five-speed
Capacity: 1,482cc
Bore & Stroke: 68.1 × 50.8mm
Maximum Speed: 257km/h (160mph)

Chassis: tubular spaceframe. Independent front and rear suspension by double wishbones and coil springs

Dimensions: wheelbase 228cm (89.75in)

This model appeared experimentally in the 1961 Italian GP and won the Manufacturers' Championship in 1962 after some excellent racing. It was the sleekest and neatest car the Brabham firm ever built, and continued its success story throughout the years of 1½-litre Formula racing. A 2-litre version did extremely well in the Australian Tasman series up to 1966.

LOTUS Type 25

Country of Origin: Great Britain
Date: 1962
Engine: Coventry-Climax, rear mounted, V-8; inclined ohvs; two ohcs per bank; downdraught carburettors
Gears: five-speed
Capacity: 1,495cc
Bore & Stroke: 63 × 60mm
Maximum Speed: 257km/h (160mph)
Chassis: aluminium monocoque. Front suspension by lower wishbone, upper rocker arm, inboard coil spring. Rear suspension by lower wishbone, upper transverse link, radius rods and coil spring
Dimensions: wheelbase 231cm (91in)

The Lotus 25 monocoque chassis pioneered a fashion in racing car design which is still prevalent. It is light and easy to make and made an immediate impact when Jim Clark won the Belgian, British and American GPs. These victories gave Lotus second place in the Manufacturers' Championship for the third year running.

PORSCHE Formula 1 Type 804

Country of Origin: Germany
Date: 1962
Engine: Porsche (air-cooled); eight cylinders, horizontally opposed; inclined ohvs with two camshafts per bank; downdraught carburettors; 180bhp at 9,200rpm
Gears: six-speed baulk-ring synchromesh
Capacity: 1,492cc
Bore & Stroke: 66 × 54.5mm
Maximum Speed: 257km/h (160mph)
Chassis: tubular spaceframe. Independent front and rear suspension by double wishbones and torsion bars
Dimensions: wheelbase 230cm (90.5in)
Brakes: Porsche disc

It took Porsche a very long time to develop the engine, but even so, it was less powerful than contemporary British V-8s. It made its first appearance in the 1962 Dutch GP, where there were handling problems which needed correction. In the French GP British cars retired from the race and Porsche were able to win, but this was its only real success, the next best placing being third in the German GP. Porsche withdrew from GP racing in 1963 and concentrated on sports-car and GT racing because of the lack of possibilities for technical development in single-seater racing.

ATS Tipo 100 Formula 1

Country of Origin: Italy
Date: 1963
Engine: ATS, V-8; Inclined ohv with two ohcs per bank; Lucas fuel-injection; 190bhp at 10,000rpm
Gears: six-speed ATS-Colotti
Capacity: 1,494cc
Bore & Stroke: 66 × 54.6mm
Maximum Speed: 233km/h (145mph)
Chassis: tubular spaceframe. Independent front suspension by wishbone and rocker arm with inboard coil spring and damper unit. Independent rear suspension by wishbones, links and coil springs
Dimensions: wheelbase 232cm (91.5in)

This car was a failed attempt to beat Enzo Ferrari at his own game. The company which produced it was made up of several ex-Ferrari top personnel, but was driven with dissent from the start. Hill and Baghetti drove the cars in the Belgian GP, but they did badly. A trip to America for the Watkins Glen and Mexico City races was also a failure.

BRABHAM BT20

Country of Origin: Great Britain
Date: 1966
Engine: Repco V-8; two valves per cylinder; single ohc; 315bhp at 7,250rpm; fuel injection
Capacity: 2,995cc
Bore & Stroke: 88.9 × 60.3mm
Chassis: tubular spaceframe. Independent suspension front and rear by wishbones, links and coil springs
Brakes: disc

This model was an immediate success, giving Brabham his third World Championship title in 1966 and the Manufacturer's Championship. In 1967 Denny Hulme won the Drivers' Championship and the Manufacturers' Championship again.

COOPER-MASERATI T81

Country of Origin: Great Britain/Italy
Date: 1966
Engine: Maserati V-12; two valves per cylinder; two ohcs; 320bhp at 9,500rpm; fuel injection
Capacity: 2,989cc
Bore & Stroke: 70.4 × 64mm
Chassis: stressed-skin monocoque. Independent front and rear suspension by wishbone rocker arm and links with coil springs inboard at front
Brakes: disc

The T81s were large, heavy cars with a chassis designed by Tony Robinson. They were bought by customers who included Jo Bonnier and Guy Ligier. Surtees gave Cooper their first major success for four years by winning the Mexican GP. Rodriguez drove a T81 to victory in the South African GP in 1967. This car was the last of Cooper's Formula 1 successes.

EAGLE T2G Formula 1

Country of Origin: Great Britain/USA
Date: 1967
Engine: Gurney Weslake, V-12, 3-litre; inclined ohvs, four per cylinder, with two ohcs per bank; Lucas fuel-injection
Gears: five-speed Hewland
Capacity: 2,997cc
Bore & Stroke: 72.8 × 60.3mm
Maximum Speed: 298km/h (185mph)
Chassis: aluminium monocoque. Independent front suspension by lower wishbone and upper rocker arm operating inboard coil spring. Independent rear suspension by lower reversed wishbone, transverse link, radius rods and coil spring
Dimensions: wheelbase 245cm (96.5in)

This model was developed and built by Dan Gurney for the new 3-litre GP Formula of 1966 and for Indianapolis in America. However, the V-12 engine, built by Harry Weslake Research at Rye in England, was slow in arriving. In the interim, Gurney used the T2G F1 chassis with a 2.75-litre Climax FPF four-cylinder engine, and came fifth in the 1966 French GP and fifth at Mexico City. The V-12 finally arrived and Gurney won the 1967 Belgian GP. The V-12s were beset with problems throughout their career, and in 1969 Gurney's backers dropped out of Formula 1 racing altogether.

FORD GT40 MK4

Country of Origin: USA
Date: 1967
Engine: GT40, 90° V-8; pushrod ohvs; twin Holley carburettors; 500bhp at 5,000rpm
Gears: five-speed manual
Capacity: 4,736cc
Bore & Stroke: 101.6 × 72.9mm
Chassis: unitary aluminium honeycomb monocoque. Independent front suspension by double wishbones, coil springs and hydraulic coaxial dampers. Independent rear suspension by single transverse top link, lower wishbones, twin trailing arms and hydraulic coaxial dampers

Dimensions: wheelbase 241cm (95in). Track 140cm (55in)
Brakes: four-wheel disc

The Mark 4 was the last of the GT40 series before Ford finally withdrew from racing. It was an outstanding sports-racing model with a reliable engine. Dan Gurney and A. J. Foyt drove one in the 1967 Le Mans to gain Ford its second victory at 218km/h (135.5mph).

HONDA RA300 Formula 1

Country of Origin: Japan
Date: 1967
Engine: Honda, V-12, air-cooled; four valves per cylinder; four ohcs; fuel injection; 400bhp at 10,500rpm
Gears: five-speed

Capacity: 2,992cc
Bore & Stroke: 78 × 52.2mm
Maximum Speed: 282km/h (175mph)
Chassis: aluminium monocoque. Front suspension by upper rocker arm operating inboard coil spring, lower wishbone. Rear suspension by single top link, lower wishbone, radius rods, coil spring

Dimensions: wheelbase 241cm (94.75in)
The RA300 was a very long and heavy car, and in Formula 1 racing this was a great disadvantage since increased horsepower did not make up for it. British cars therefore retained superiority, being smaller and lighter, despite having less horsepower. John Surtees got Eric Broadley of Lola to design a new monocoque chassis and suspension to take the enormous engine and after an epic race with Jack Brabham won the 1967 Italian GP; it also came fourth at Mexico.

LOTUS 49

Country of Origin: Great Britain
Date: 1967
Engine: Ford-Cosworth V-8; four valves per cylinder; two ohcs per bank; 430bhp at 9,000rpm; fuel injection
Gears: ZF gearbox
Capacity: 2,993cc
Bore & Stroke: 85.6 × 64.8mm
Chassis: forward monocoque, stressed engine. Independent front suspension with inboard coil springs. Independent rear suspension with coil springs
Brakes: disc

1967 marked a turning point in motor racing history with the advent of the Ford-Cosworth DFV V-8 engine. Initially the engine was available only to Lotus, and the Lotus 49 was the first model to use it. It made its debut in Holland driven by Graham Hill and Jim Clark, who scored a spectacular win after Hill retired with cam-drive failure. Clark won three more events that season, but chassis, engine and gearbox failures robbed him of the World Championship. 1968 proved another successful GP season with wins by Clark, Hill, Stewart, Hulme and McLaren. The 49 was phased out in 1970.

STP-Turbocar

Country of Origin: USA
Date: 1967
Engine: Pratt & Whitney gas turbine ST6B; fuel injection
Capacity: equivalent to 4.2 litres
Maximum Speed: 290km/h (180mph)

Chassis: fabricated steel frame. Independent front suspension by wishbones and coil springs. Independent rear suspension by transverse links and coil springs

Dimensions: wheelbase 244cm (96in)

The power plant was mounted to the left of the chassis, with the driver on the right, there was a duct on the nose to take air to the turbine. It had four-wheel drive. In the 1967 Indianapolis 500-Mile Race the Turbocar was in the lead until near the end when it had to retire due to a bearing malfunction. The Turbocar was so successful that cars of its type were eventually banned from USAC racing to protect the interests of those involved with the piston engine.

McLAREN-FORD M7

Country of Origin: Great Britain
Date: 1968
Engine: Cosworth-Ford, V-8; four valves per cylinder; two ohcs; 430bhp at 9,000rpm; fuel injection
Capacity: 2,993cc
Bore & Stroke: 85.6 × 64.8mm
Chassis: forward monocoque with stressed engine. Independent suspension front and rear by wishbones, links and coil springs, outboard all-round
Brakes: disc

The M7 series was introduced in 1968. Bruce McLaren and Denny Hulme drove M7s to victory in their first season, including the Belgian and Canadian GPs, and McLaren won the Race of Champions in the same year. They were raced in 1969, but were superseded by the four-wheel drive M9A car.

BMW Formula 2

Country of Origin: Germany
Date: 1969
Engine: BMW-M20, four-cylinder, 1.6-litre; four valves per cylinder with two ohcs; 225bhp at 10,300rpm
Gears: five-speed Hewland
Capacity: 1,596cc
Bore & Stroke: 89 × 64mm
Maximum Speed: 257km/h (160mph)
Chassis: aluminium monocoque with tubular engine bay (Lola T102). Front suspension by double wishbones and coil springs. Rear suspension by transverse links, radius rods and coil springs
Dimensions: wheelbase 209cm (82.25in)

Although BMW had been involved in sports-car racing for many years, it was only in 1966 that the first single-seater racing car was produced. The engine was initially mounted in a Lola chassis, but after one season, the monocoque chassis, designed and built by Dornier, became available. These cars were quite successful in Formula 2 racing during 1970. After that, BMW withdrew from single-seater racing.

CHEVRON B16

Country of Origin: Great Britain
Date: 1969
Engine: Ford Cosworth FVC
Gears: five-speed Hewland
Capacity: 1,800cc
Chassis: round and square steel tubing, monocoque. Independent front suspension by upper forward mounted radius arms and locating link, tower triangulated wishbone, shock absorbers with coil springs, anti-roll bar. Independent rear suspension by upper and lower forward mounted radius arms, upper link and lower triangulated wishbone, adjustable shock absorbers with coil springs.
Dimensions: wheelbase 234cm (93in). Track (front) 132cm (52in); track (rear) 132cm (52in)

Steering: rack-and-pinion
Brakes: disc
Body: lightweight glass-reinforced plastic

The B16 is generally agreed to be one of the most beautiful sports racing cars ever built. Its first win was at the Nurburgring in the 1969 500-km Race, driven by Brian Redman. Its first British win was at Snetterton on Good Friday 1970. Three days later, Brian Redman was only just beaten by a 4.5-litre Porsche in the European Championship Race at Thruxton. It won the 1970 Nurburgring 500-km Race and at Spa-Francochamps in Belgium.

CROSSLÉ 16F Formula Ford

Country of Origin: Ireland
Date: 1969
Engine: Ford, four-cylinder; pushrod ohv; single carburettor
Gears: four-speed
Capacity: 1,598cc
Bore & Stroke: 81 × 77.6mm
Maximum Speed: 209km/h (130mph)
Chassis: tubular spaceframe. Independent front suspension by wishbones and coil springs. Independent rear suspension by links, radius rods and coil springs
Dimensions: wheelbase 223cm (88in)

John Crosslé began building his own cars in 1957 and won the Ford Championship of Ireland in 1958, 1959 and 1960. Crosslé cars also won the Ford Points Championship in 1961, 1962 and 1963. By 1967 Crosslé was constructing about 24 single-seater cars every year and the 16F model, developed with C. T. Wooler Ltd, was very successful. Gerry Birrell drove it to win the 1969 European Formula Ford Championship. The 16F, like all Crosslé's cars had sound engineering and was perfectly finished.

LOLA T70 Mk 3B

Country of Origin: Great Britain
Date: 1969
Engine: Chevrolet 90° V-8; pushrod ohv; four twin choke Weber carburettors; 460bhp at 6,300rpm
Gears: four-speed Hewland manual or five-SPEED ZF manual
Capacity: 5,463cc
Bore & Stroke: 81.9 × 102.7mm
Chassis: monocoque. Independent front suspension by upper wishbones, lower links, coil springs, dampers and anti-roll bar. Independent rear suspension by

radius arms, lower wishbones, upper links, coil springs and dampers
Dimensions: wheelbase 241cm (95in); track 137cm (54in)
Brakes: four-wheel disc
This model had an immediate major racing success, coming both first and second in the Daytona 24-Hour race. It had a string of other successes in both 1969 and 1970, becoming the most successful car ever in its class. Despite this, production was stopped in 1970.

LOTUS 63

Country of Origin: Great Britain
Date: 1969
Engine: Ford-Cosworth V-8; four valves per cylinder; two ohcs; 430bhp at 9,000rpm; fuel injection
Capacity: 2,993cc
Bore & Stroke: 85.6 × 64.8mm
Chassis: forward monocoque with reversed stressed engine mounting. Independent front suspension by inboard coil springs. Independent rear suspension by inboard coil springs. Four-wheel drive
Brakes: disc

The Lotus 63 was one of several four-wheel drive cars which appeared in 1969. It was a difficult car to control and gradually more and more power was distributed to the rear wheels until it very nearly became a rear-wheel drive. Only two were built and the best result was by Rindt who came second in the 1969 Oulton Park Gold Cup non-championship race.

MATRA-COSWORTH MS80 Formula 1

Country of Origin: Great Britain/France
Date: 1969
Engine: Cosworth DFV, V-8; four valves per cylinder; two ohcs per bank; Lucas fuel injection
Gears: five-speed Hewland
Capacity: 2,993cc
Bore & Stroke: 85.6 × 64.8mm
Maximum Speed: 290km/h (180mph)

Chassis: aluminium monocoque with engine a stressed member. Front suspension by double wishbones and coil springs. Rear suspension by upper wishbone, lower links, radius rods and coil springs
Dimensions: wheelbase 240cm (94.5in)
Driven by Jackie Stewart, the MS80 won six Grand Prix to clinch the World Championship in 1969, and gave Matra the Manufacturers' title.

DE TOMASO Type 38 Formula 1

Country of Origin: Italy
Date: 1970
Engine: Cosworth DFV, V-8; four valves per cylinder, four ohcs; Lucas fuel-injection
Gears: five-speed Hewland
Capacity: 2,993cc
Bore & Stroke: 85.6 × 64.8mm
Maximum Speed: 282km/h (175mph)

Chassis: aluminium monocoque with cast-magnesium centre bulkhead, with engine a stressed member. Independent front suspension by wishbones and coil springs. Rear suspension, independent by transverse link, wishbone, radius rods and coil spring
Dimensions: wheelbase 241cm (95in)
The Argentinian racing driver De Tomaso, having retired from racing, took to constructing racing cars for amateur drivers. In 1970 however, he produced Formula 1 cars for the Frank Williams racing team.

FERRARI 312B

Country of Origin: Italy
Date: 1970
Engine: Ferrari 12-cylinder horizontally opposed; four valves per cylinder; two ohcs; 460bhp at 12,000rpm; fuel injection
Capacity: 2,998cc
Bore & Stroke: 79 × 52.8mm
Chassis: tubular spaceframe with stressed panelling. Independent suspension front and rear with coil springs, inboard at front
Brakes: disc

The first two appearances of this car were unsuccessful despite being driven by Jacky Ickx. Ickx nearly lost his life in a collision in Spain when his car was destroyed by fire. Ignazio Giunti finally obtained championship points in the 312B by coming fourth in the Belgian GP. At Zandvoort Regazzoni came fourth, and at Hockenheim Ickx's 312B was only narrowly beaten by Rindt. It was in Austria that Ickx finally brought the 312B in first closely followed by Regazzoni, and Ickx again won in Canada and Mexico City.

LOTUS Type 72 Formula 1

Country of Origin: Great Britain
Date: 1970
Engine: Ford-Cosworth, V-8; four valves per cylinder, two ohcs per bank; Lucas fuel injection
Gears: five-speed Hewland
Capacity: 2,993cc
Bore & Stroke: 85.6 × 64.8mm
Maximum Speed: 298km/h (185mph)
Chassis: aluminium monocoque, with engine a stressed member. Front suspension, wishbones and torsion bars. Rear suspension, wishbones, radius rods and torsion bars

Dimensions: wheelbase 254cm (100in)
Brakes: disc, inboard

Another in the long line of Lotus innovations, the 72 had torsion bar suspension in place of coil springs, and with the disc brakes mounted inboard. Jochen Rindt won four successive races in the 72, starting with the 1970 Spanish GP, before his career was ended in a crash during a practice run for the Italian GP. Fittipaldi took Rindt's place, winning the United States GP, and the Lotus 72 gave Lotus the 1970 F1 Manufacturers' Cup.

MATRA MS 120 Formula 1

Country of Origin: France
Date: 1970
Engine: Matra, V-12, 3-litre; four valves per cylinder; two ohcs per bank; Lucas fuel injection
Gears: five-speed Hewland
Capacity: 3,000cc
Bore & Stroke: 79.7 × 50mm
Maximum Speed: 282km/h (175mph)
Chassis: aluminium monocoque. Front suspension by double wishbones and coil springs. Rear suspension

by transverse links, radius rods and coil springs
Dimensions: wheelbase 243cm (95.75in)
Despite its inelegant appearance this car had a first-class chassis and good roadholding. The engine, however, was the least powerful used by cars of its type and class, and the best it could do was third place at Monaco, Monza and Spa in 1970.

FERRARI 312P

Country of Origin: Italy
Date: 1972
Engine: flat-12; two ohcs per bank; Lucas fuel injection; 400bhp at 10,800rpm
Gears: five-speed manual
Capacity: 2,991cc
Bore & Stroke: 78.5 × 52.5mm
Chassis: tubular and sheet monocoque. Independent front suspension by wishbones, coil springs, dampers and anti-roll bar. Independent rear suspension by radius arms, wishbones, coil springs, dampers and anti-roll bar
Dimensions: wheelbase 222cm (87.5in). Track (front) 142cm (56in); track (rear) 140cm (55in)
Brakes: four-wheel disc

The 312P dominated the 1972 World Championship of Makes when it won ten races out of ten entered, including eight one-two results. Regular drivers were Jacky Ickx, Mario Andretti, Ronnie Peterson, Brian Redman and Tim Schenken. The car was withdrawn from the Le Mans 24-Hour race because the engine was suspect, but, because of the previous winning streak, won the Championship anyway.

SURTEES TS10 Formula 2

Country of Origin: Great Britain
Date: 1972
Engine: Ford BDA. four-cylinder, 2-litre; inclined ohvs with twin ohcs; Lucas fuel injection
Gears: five-speed Hewland FG400
Capacity: 1,994cc
Bore & Stroke: 90.4 × 77.6mm
Maximum Speed: 257cm (160mph)
Chassis: aluminium monocoque. Independent front suspension by wishbones and coil springs. Independent rear suspension by transverse link, wishbone and twin radius rods, coil springs

Dimensions: wheelbase 241cm (95in)

John Surtees won seven motorcycle world championships before taking to cars. He joined Ferrari in 1963 and won the World Driver's Championship in 1964. In 1969 he began to run his own racing team driving his own cars. These have been more successful in Formula 2 than in Formula 1, Mike Hailwood winning the European F2 Championship in 1972 in this model.

TECNO PA123 Formula 1

Country of Origin: Italy
Date: 1972
Engine: Pederzani, 12 cylinders horizontally opposed; inclined, ohvs with two camshafts per bank; Lucas fuel injection
Gears: five-speed Hewland FG400
Capacity: 2,995cc
Bore & Stroke: 80.98 × 48.46mm

Maximum Speed: 290km/h (180mph)
Chassis: aluminium monocoque. Independent front suspension by double wishbones and inboard coil springs. Independent rear suspension by single transverse top link, reversed lower wishbone, double radius rods and coil springs
Dimensions: wheelbase 247cm (97.25in)
The Pederzani brothers started out by building Karts

then progressed through Formula 3 and Formula 2 until in 1971, backed by Martini finance, they were able to start work on a Formula 1 racing car, the PA123. It went through several major design modifications but despite this intensive development work scored no successes. The racing team lacked direction and after many disagreements the whole project was abandoned.

PORSCHE 917–30

Country of Origin: Germany
Date: 1973
Engine: flat-8, twin turbo; 5.4-litre; air-cooled; fuel injection; 1,100bhp at 7,800rpm
Gears: four-speed
Capacity: 5,374cc
Chassis: tube frame. Front suspension by wishbone and thrust rod. Rear suspension by wishbone and trailing arm
Dimensions: wheelbase 250cm (98.5in). Track (front) 167cm (65.8in); track (rear) 156.5cm (61.6in)
Steering: rack-and-pinion
Brakes: ventilated disc

Described by one eminent racing authority as 'the ultimate overkill Porsche', the 917–30 was really an improved 917–10, the previous year's CanAm winner, developed to challenge the consistently victorious McLarens. The 917–30's first couple of races were disappointing, but in subsequent events, the car showed its paces by winning six consecutive CanAm races.

TYRRELL 006/2

Country of Origin: Great Britain
Date: 1973
Engine: Ford-Cosworth V-8; four valves per cylinder; two ohcs; fuel injection; 460bhp at 10,000rpm
Capacity: 2,993cc
Bore and Stroke: 85.6 × 64.8mm
Chassis: Forward stressed-skin monocoque with stressed engine. Independent front and rear suspension by coil springs

Brakes: disc
This car was built for Jackie Stewart and made its first appearance in the non-Championship International Trophy race at Silverstone. Despite snow and a slow puncture, Stewart won and then went on to win the Belgian, Dutch, German and Monaco GPs.

HILL GH1

Country of Origin: Great Britain
Date: 1975
Engine: Cosworth DFV, V-8; four valves per cylinder, two ohcs per bank; Lucas fuel-injection
Gears: five-speed Hewland FGA400
Capacity: 2,993cc
Bore & Stroke: 85.6 × 64.8mm
Maximum Speed: 282km/h (175mph)
Chassis: aluminium monocoque, engine a stressed member. Independent front suspension by double wishbones and coil springs. Independent rear suspension by single transverse top link, twin lower links, double radius rods, coil springs
Dimensions: wheelbase 256cm (101in)

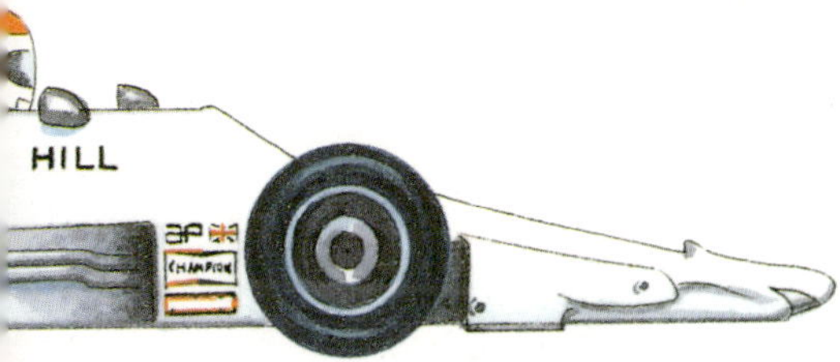

In 1973 Graham Hill set up an organisation to run his own car. Having obtained financial backing he bought the parts for a Shadow DN1 and assembled them himself, using a Cosworth Ford DFV engine and Hewland gearbox. The following year Lola supplied the parts of their F5000 design for Hill to construct as Formula 1. With Tony Brise as a driver, Hill had a very promising combination, but tragically, the whole team – Hill, Brise, designer and mechanics – were killed when their plane crashed near Elstree in November 1975.

PARNELLI VPJ4
Country of Origin: Great Britain/USA
Date: 1975
Engine: Cosworth DFV, V-8; four valves per cylinder, with two ohcs per bank; Lucas fuel injection
Gears: five-speed Hewland
Capacity: 2,993cc
Bore & Stroke: 85.6 × 64.8mm
Maximum Speed: 282km/h (175mph)

Chassis: aluminium monocoque, with engine a stressed member. Independent front suspension by wishbones and torsion bars. Independent rear suspension by transverse links, radius rods and torsion bars
Dimensions: wheelbase 254cm (100in)
Brakes: Girling
Rufus Parnelli Jones, who won the 1963 Indianapolis,

became involved in Formula 1 racing as a constructor. His designer, Maurice Phillippe, had previously worked for Lotus, so Parnelli cars bore close resemblances to Lotus cars. The VPJ4 itself, was similar to an uprated Lotus 72. Its first race was the 1974 Canadian Grand Prix where it came seventh. it was not successful in subsequent races, and quietly disappeared early in 1976.

SHADOW DN5

Country of Origin: Great Britain/USA
Date: 1975
Engine: Cosworth DFV, V-8; four valves per cylinder, two ohcs per bank; Lucas fuel injection
Gears: five-speed Hewland FL200
Capacity: 2,993cc
Bore & Stroke: 85.6 × 64.8mm
Maximum Speed: 282km/h (175mph)
Chassis: aluminium monocoque with engine a stressed member. Independent front suspension by

double wishbones and coil springs. Independent rear suspension by transverse link, wishbone, radius rods and coil springs
Dimensions: wheelbase 267cm (105in)
The Shadow was designed by Tommy Southgate, who had previously worked for Lola, Brabham, Eagle and BRM, and built by AVS with financial backing by Universal Oil Products of Illinois. In 1975 the firm considered using the Matra V-12 engine, but only one car (the DN7) was built.

ENSIGN F1-N176

Country of Origin: Great Britain
Date: 1976
Engine: Cosworth DFV, V-8; four valves per cylinder, two ohcs per bank; Lucas fuel-injection

Gears: five-speed Hewland
Capacity: 2,993cc
Bore & Stroke: 85.6 × 64.8mm
Maximum Speed: 282km/h (175mph)
Chassis: aluminium monocoque with engine a stressed member. Independent front suspension by double wishbones and coil springs. Independent rear suspension by single transverse top link, twin

lower links, double radius rods and coil spring
Dimensions: wheelbase 261cm (102.75in)

Although it did not win a race outright, the Ensign F1-N176 was certainly competitive in Formula 1 racing. It had a sound and uncomplicated design, a special feature being the front brakes with calipers fore and aft of each disc.

FITTIPALDI FD/04

Country of Origin: Great Britain/Brazil
Date: 1976
Engine: Cosworth DFV, V-8; four valves per cylinder, two ohcs per bank; Lucas fuel-injection
Gears: five-speed, Hewland
Capacity: 2,993cc
Bore & Stroke: 85.6 × 64.8mm
Maximum Speed: 282km/h (175mph)
Chassis: aluminium monocoque, with engine a stressed member. Independent front suspension by double wishbones, coil springs. Independent rear suspension by single transverse top link, twin lower links, radius rods and coil springs
Dimensions: wheelbase 243cm (95.75in)

The Brazilian Fittipaldi brothers, Emerson and Wilson both became Formula 1 drivers; but while Emerson went on to win the World Championship in 1972, Wilson began to build Formula 1 cars. With Richard

Divila and finance from the Copersucar company he built the first Fittipaldi car, FD/01. This was smashed up in its first race and the second and third cars were not much more successful. The 1976 model was to be driven by Emerson, who had left the McLaren team for this reason. It began well in the Brazilian GP, but made little progress after that.

MARTINI MK19 Formula 2

Country of Origin: France
Date: 1976
Engine: Renault, V-6; four valves per cylinder; four ohcs; fuel injection
Gears: five-speed, Hewland FG200
Capacity: 1,997cc
Bore & Stroke: 86 × 57.3mm
Maximum Speed: 257km/h (160mph)
Chassis: monocoque. Independent front suspension by double wishbones and coil springs. Independent

rear suspension by single transverse top link, twin lower links, double radius rods and coil springs
Dimensions: wheelbase 241cm (95in)

Renato (Tico) Martini was an entirely self-taught car constructor and had build over 200 cars by 1976. In that year a MK19 was runner-up in the Formula 2 European Championship. The MK serial letters stand for Martini-Knight (not an abbreviation of Mark), Knight being the name of a family who helped Martini to become established.

185

TYRRELL P34

Country of Origin: Great Britain
Date: 1976
Engine: Cosworth DFV, V-8; four valves per cylinder; two ohcs per bank; Lucas fuel injection
Gears: five-speed Hewland
Capacity: 2,993cc
Bore & Stroke: 85.6 × 64.8mm
Maximum Speed: 290km/h (180mph)
Chassis: aluminium monocoque with engine a stressed member. Independent front suspension by double wishbones and coil springs. Independent rear suspension by single top link, twin lower links, radius rods and coil springs

Dimensions: wheelbase to front axle 254cm (100in). To centre axle 209.5cm (82.5in)

One of the most unusual features of this model was that it had six wheels. Project 34 was a prototype built in 1975, a research model to illustrate the basic concept, and by 1976 the Tyrrell team were racing 34/2, 34/3 and 34/4. By the middle of the 1976 season Scheckter won the team's first victory in the new model by achieving third place in the Manufacturer's Championship.

RENAULT RS01 Formula 1

Country of Origin: France
Date: 1977
Engine: Renault-Gordini, V-6; four valves per cylinder, two ohcs per bank; exhaust driven turbocharger
Gears: six-speed, Hewland
Capacity: 1,492cc
Bore & Stroke: 86 × 42.8mm
Maximum Speed: 306km/h (190mph)

Chassis: aluminium monocoque, with engine a stressed member. Front suspension by double wishbones and inboard coil springs. Rear suspension by twin lower links, single upper link, double radius rods, coil springs

Dimensions: wheelbase 250cm (98.5in)

This model broke with normal Formula 1 practice by using the turbo-charged, 1½-litre engine; other Formula 1 makes used unsupercharged 3-litre engines. The engine had a potential of 5,000bhp on a 1½-litre capacity.

FERRARI 312 T – 4

Country of origin: Italy
Date: 1979
Engine: Ferrari 312 Boxer; Flat-12; 510bhp at 12,000rpm. Lucas fuel injection
Gears: Ferrari 015 Trasversale five-speed
Capacity: 2,991cc
Bore & Stroke: 78.5 × 51.5mm
Chassis: Suspension (front) Double wishbones,

inboard springs; suspension (rear) Bottom wishbones, single top links, inboard springs. Suspension dampers Koni
Dimensions: wheelbase 270cm (106.3in). Track (front) 170cm (67in); track (rear) 160cm (63in)
Tyres: Michelin
Steering: Ferrari rack and pinion
Brakes: Lockheed

WILLIAMS FW06-FORD

Country of origin: Great Britain
Date: 1979
Engine: Ford–Cosworth DFV; V-8; 470bhp (minimum) at 10,800rpm. Lucas fuel injection
Gears: Hewland FGA400 six-speed
Capacity: 2,993cc
Bore & Stroke: 85.6 × 64.8mm
Chassis: Suspension (front) Top rocker arms, bottom

wishbones, inboard springs; suspension (rear) lower wishbones, top rocker arms and links, inboard springs. Suspension dampers Koni
Dimensions: wheelbase 269cm (106in). Track (front) 172cm (68in) Track (rear) 160cm (63in)
Tyres: Goodyear
Steering: Williams rack and pinion
Brakes: Lockheed

LOTUS 81 – COSWORTH

Country of origin: Great Britain
Date: 1980
Engine: Ford–Cosworth DFV; V-8; 470bhp (minimum) at 10,800rpm. Lucas fuel injection
Gears: Hewland/Lotus five-speed
Capacity: 2,993cc
Bore & Stroke: 85.6 × 64.8mm
Chassis: Suspension (front and rear) Top rocker arms,

bottom wishbones, inboard springs. Suspension dampers Koni
Dimensions: wheelbase 274cm (108in). Track (front) 177cm (70in); track (rear) 162cm (64in)
Tyres: Goodyear
Steering: Knight rack and pinion
Brakes: Lockheed

PORSCHE 956C

Country of Origin: Germany
Date: 1981
Engine: Porsche, six-cylinder horizontally opposed, with water-cooled cylinder heads; four valves per cylinder; four ohcs with open wheel drive; twin KKK K26 turbochargers at 1.2bar; 630bhp at 8,200rpm
Gears: five-speed fully synchronised Porsche with air cooling, or electronic double-clutch Porsche five-speed
Capacity: 2,649cc
Bore & Stroke: 92.3 × 66mm
Maximum Speed: 350km/h (220mph)

Chassis: aluminium monocoque. Independent suspension both front and rear with double wishbones, titanium coil springs, Bilstein shock absorbers
Dimensions: wheelbase 265cm (104in). Track (front) 166.5cm (65.5in); track (rear) 154.5cm (60.8in)
Brakes: twin circuit Porsche with ventilated discs and twin calipers per wheel
The 956C scored a hat-trick at its debut at Le Mans in 1982. It won Porsche its ninth World Championship of Makes, and Jacky Ickx the World Championship Driver's title. The 956C is not just a successful racing car, it is, as Porsche put it, a 'laboratory on wheels', used to test many components which will eventually be used in production models. It won the 'Motor Sport Vehicle of the Year' title in 1982.

ATS D5-COSWORTH

Country of Origin: Italy/Great Britain
Date: 1982
Engine: Ford–Cosworth DFV; V-8; 480bhp at 11,100rpm
Gears: ATS/Hewland five/six-speed
Capacity: 2,993cc
Bore & Stroke: 85.6 × 64.8mm
Chassis: Suspension (front and rear) Top rocker arms,

lower wishbones, inboard springs. Suspension dampers Koni
Dimensions: wheelbase 266cm (105in). Track (front) 175cm (69in); track (rear) 162cm (64in)
Tyres: Avon/Michelin
Steering: ATS
Brakes: Lockheed

MARCH 821 COSWORTH

Country of Origin: Great Britain
Date: 1982
Engine: Ford–Cosworth DFV; V-8; 480bhp at 11,100rpm
Gears: five-speed Hewland
Capacity: 2,993cc
Bore & Stroke: 85.6 × 64.8mm

Chassis: Suspension (front and rear) Top rocker arms, lower wishbones, inboard springs. Suspension dampers Koni
Dimensions: wheelbase 269cm (106in). Track (front) 172cm (68in); track (rear) 167cm (66in)
Tyres: Pirelli/Avon Michelin
Steering: March
Brakes: Lockheed

LANCIA LC2/83

Country of Origin: Italy
Date: 1983
Engine: LC268/T 90°; V-8; two KKK turbochargers; Weber-Magneti Marelli indirect electronic injection; 720bhp

Gears: five-speed, Abarth
Capacity: 3,014.7cc
Bore & Stroke: 84 × 93mm
Chassis: Independent suspension front and rear by wishbones and Bilstein air-hydraulic dampers
Dimensions: wheelbase 474.7cm (187in). Track (front) 147cm (58in); track (rear) 134cm (53in)
Brakes: Brembo
One of the competitors in the 1983 Le Mans 24-Hour Race.

MAZDA 717C

Country of Origin: Japan
Date: 1983
Engine: Mazda rotary, 2.6-litres; Bosch fuel injection; 300bhp at 9,000rpm; rear mounted
Gears: five-speed Hewland FG
Maximum Speed: 298km/h (185mph)
Chassis: monocoque tube of honeycomb aluminium with steel plating for suspension pick-up mounting points, inboard suspension with coil spring damper units
Brakes: disc, front and rear, with AP four-piston calipers

Body: Kevlar carbon-fibre

The Mazda Group C race cars, designed and run by the factory team under the name Mazdaspeed, are part of a long-term plan to compete in World Endurance Championship events. For the last two years the team has entered the Silverstone Six-Hours, Le Mans 24-Hours and the Fuji 1,000-kilometre races. Their greatest success was at Le Mans in 1983 where they took first and second places in the Junior Group C category.

ARROWS A7

Country of Origin: Great Britain
Date: 1984
Engine: BMW four-cylinder 1.5-litre turbo; 750bhp at 10,500rpm on full boost
Gears: Six-speed Arrows with Hewland internals
Chassis: aluminium and carbon honeycomb construction. Front suspension, rising rate pull rod type, upper and lower wishbones, inboard spring operated by pull rod connected to upper wishbone. Rear suspension by top rocker, inboard springs and lower wishbone

Dimensions: wheelbase 267cm (105in). Track (front) 183cm (72in); track (rear) 163cm (64in)
Steering: rack-and-pinion
Brakes: AP balanced braking rear. Brembo single caliper front. AP master cylinder
Body: carbon fibre/Kevlar composite

The 1984 season was a disappointing one for the Arrows/BMW combination. The A7 was bugged by engine problems on a number of occasions, but came sixth in the Brazilian GP, fifth at San Marino and fifth in the Austrian GP.

BRABHAM BT53

Country of Origin: Great Britain
Date: 1984
Engine: BMW four-cylinder, in line; water-cooled; two ohcs; four ohvs per cylinder; 640bhp at 10,500rpm; exhaust gas turbocharger by KKK
Capacity: 1,499cc
Bore & Stroke: 89.2 × 60mm
Chassis: front suspension by double wishbones and pushrods. Rear suspension by double wishbones and pushrods; Koni suspension dampers
Dimensions: wheelbase 292cm (115in). Track (front) 175cm (69in); track (rear) 165cm (65in)

Tyres: Michelin
Steering: rack-and-pinion
Brakes: Hitco/Brabham/Girling/AP

This car was the only challenger to the magnificent McLaren MP4/2, having enormous power. But the BMW engine proved to be unreliable and Piquet was forced to retire from nearly half the season's races. However he had victories in the Canadian and Detroit GPs and came second in the Austrian GP.

FERRARI 126C4 Turbo

Country of Origin: Italy
Date: 1984
Engine: Ferrari 126C4; two KKK superchargers; 680bhp at 11,500rpm; Lucas/Ferrari-Weber/Marelli fuel injection
Gears: five-speed Ferrari
Capacity: 1,496cc
Bore & Stroke: 81 × 48.4mm
Chassis: front suspension by double wishbones and pull rods. Rear suspension by double wishbones and coil rods; Koni suspension dampers

210

Dimensions: wheelbase 260cm (102.4in). Track (front) 179cm (70.5in); track (rear) 164.4cm (64.7in)
Tyres: Goodyear radials
Steering: rack-and-pinion
Brakes: Brembo/SEP

1984 was a troublesome season for Ferrari, with the C4s needing alterations almost every time they raced. Ferrari achieved only one victory in 1984, at Zolder in the Belgian GP. The C4s also came second at Imola, Dallas, Monza and at the Nurburgring.

LIGIER JS23

Country of Origin: France
Date: 1984
Engine: Renault EF1; V-6; Garret turbocharger; 750bhp at 11,500rpm; electronic Renix fuel injection
Gears: Ligier/Hewland
Capacity: 1,492cc
Bore & Stroke: 86 × 42.8mm
Chassis: Front suspension by upper and lower wishbones and push rods. Rear suspension by upper and

lower wishbones and pushrods; Koni suspension dampers
Dimensions: wheelbase 281cm (111in). Track (front) 180cm (71in); track (rear) 171cm (67in)
Tyres: Michelin
Brakes: Brembo
The JS23 was a disappointing car, proving very unreliable. Its best result was fifth in the 1984 South African GP, driven by Andrea de Cesaris.

LOTUS JPS 95T Formula 1

Country of Origin: Great Britain
Date: 1984
Engine: Renault V-6 turbo, 1.5-litre
Gears: Team Lotus designed and built with some Hewland internals
Chassis: carbon fibre/Kevlar skins with Nomex honeycomb monocoque. Front suspension by fabricated pull rod with inboard spring damper units and wide base lower wishbones. Cast magnesium uprights. Driver adjustable front anti-roll bar. Rear suspension by fabricated steel rockers and wishbones, with a rocker and pull rod operated inboard spring damper unit. Fabricated steel uprights
Dimensions: wheelbase 268cm (105.5in). Track (front) 181.6cm (71.5in); track (rear) 162cm (64in)
Tyres: Goodyear radials

Steering: rack-and-pinion
Brakes: outboard single caliper Brembo operating on ventilated iron discs
Body: Kevlar one piece nose, cockpit surround and tail. Carbon fibre flat underbody.

Designed by JPS Team Lotus Chief Engineer, Gerard Ducarouge, the 95T was built for the 1984 World Championship. The cars took pole position twice (in Rio and Dallas) and started from the front row on six occasions. A consistent performance throughout the season brought Elio de Angelis close to the World Championship, but engine and gearbox problems (which had also beset Nigel Mansell throughout the season) defeated this aim. However, de Angelis' overall performance gave him and JPS Team Lotus third place in their respective world championships.

McLAREN-TAG MP4/2

Country of Origin: Great Britain
Date: 1984
Engine: TAG turbo PO1 V-6; 750bhp; KKK tubochargers; Bosch electronic fuel injection
Gears: five-speed Hewland
Chassis: moulded carbon fibre/honeycomb composite. Front suspension by pushrod operating inboard auxiliary rocker, lower wishbones, inboard springs. Rear suspension by top rocker arms, lower wishbones and inboard springs; suspension dampers
Dimensions: wheelbase 276.8cm (109in). Track (front) 181.6cm (71.5in); track (rear) 167.6cm (66in)

Tyres: Michelin radial
Steering: rack-and-pinion
Brakes: calipers – McLaren twin caliper front and rear

The Formula 1 car of the 1984 season and one of the most successful chassis/engine combinations ever raced. Between them, Prost and Lauda notched up a string of victories right the way through the 1984 season to win the World Championship, an astonishing result for an engine in its first full season. The Space-age carbon fibre chassis is manufactured in the USA using the same technology as that used in the construction of the *Colombia* space shuttle.

RENAULT RE50

Country of Origin: France
Date: 1984
Engine: Renault EF1; turbo; V-6; four valves per cylinder; four ohcs; 750bhp at 11,500rpm; Garret turbocharger; electronic Renix fuel injection
Gears: five-speed Renault/Hewland
Capacity: 1,492cc
Bore & Stroke: 86 × 42.8mm
Chassis: front suspension by double wishbones and pull rods. Rear suspension by double wishbones and pull rods; De Carbon suspension dampers
Dimensions: wheelbase 268cm (105.6in). Track (front) 182cm (71.6in); track (rear) 166.7cm (65.6in)

Tyres: Michelin
Brakes: Brembo
The Renault RE50 began the 1984 season well, coming fifth in the Brazilian, third in the South African and second in the Belgian GP. After Tambay's second place in the French GP however, results were disappointing, the RE50 only achieving one more second place, at Brands Hatch, and third and fifth at Hockenheim.

TOLEMAN TG183B

Country of Origin: Great Britain
Date: 1984
Engine: Hart 415T; four cylinders in line; 600bhp at 2.1bar boost (race); Hart/BRA fuel injection, Holset turbocharger
Gears: five-speed Toleman/Hewland
Capacity: 1,496cc
Bore & Stroke: 88 × 61.5mm
Chassis: Front suspension by upper and lower wishbones, tension link operated inboard springs. Rear suspension upper and lower wishbones, tension link operated inboard springs; Koni suspension dampers

Dimensions: wheelbase 269cm (106in). Track (front) 183.5cm (72.25in); track (rear) 168cm (66in)
Tyres: Pirelli
Brakes: Lockheed

There was a lot of wrangling over uncompetitive tyres with Pirelli in 1984 which blew up into a public row at Imola. However, a superb performance by Ayrton Senna brought the TG183B into second place at Monaco. He subsequently came third in the British GP, proving that Toleman really could be a force in GP racing.

TYRRELL 012 Grand Prix

Country of Origin: Great Britain
Date: 1984
Engine: Ford-Cosworth DFY
Gears: five- or six-speed Tyrrell-Hewland
Chassis: front and rear suspension by pull rod, upper and lower wishbones. Rear suspension by pull rod and upper and lower wishbones
Dimensions: wheelbase 264cm (104in). Track (front) 165cm (65in); track (rear) 147cm (58in)
Tyres: Goodyear radials
Brakes: solid or ventilated AP racing disc, with single brake calipers at each corner
Body: carbon fibre

The 012 is the successor to the Tyrrell 011 which won the 1983 Detroit Grand Prix. The monocoque panels are a 50/50 mix of aluminium skin/honeycomb core and carbon skin/honeycomb core with bulkheads of aluminium alloy. This combination gives a structure of equal torsional stiffness to the 011, but is about 30% lighter. As a result the 012 has the 011's advantages but with a smaller, lighter chassis which is structurally and aerodynamically more efficient.

WILLIAMS FW09-HONDA

Country of Origin: Great Britain
Date: 1983
Engine: Honda RA163E, turbo; V-6; four valves per cylinder; four ohcs; 600+bhp at 11,000rpm
Gears: six-speed Williams/Hewland
Capacity: 1,497cc
Chassis: front suspension by upper and lower wishbones, pull rods, inboard spring/dampers, Rear suspension by top rocker arms, lower wishbones, inboard spring/dampers. Koni suspension dampers

Dimensions: wheelbase 266.7cm (105in). Track (front) 180.3cm (71in); track (rear) 167.6cm (66in)
Tyres: Goodyear
Steering: rack-and-pinion
Brakes: AP/SEP
The FW09-Honda made its debut at the end of 1983 and the Williams team were still getting used to the new turbo car in the 1984 season. Driven by Rosberg it won the Dallas GP, came second at Rio and fourth at Zolder and Monte Carlo.

RAM 03 – HART

Country of Origin: Great Britain
Date: 1985
Engine: Hart 415 Turbo; 4-in line; 750bhp at 10,500rpm
Gears: RAM six-speed
Capacity: 1,496cc
Bore & Stroke: 88 × 61.5mm
Chassis: Suspension (front and rear) by double wishbones and pull rods. Suspension dampers Koni

Dimensions: wheelbase 278cm (109.6in). Track (front) 181cm (71.38in); track (rear) 165cm (65in)
Tyres: Pirelli
Steering: RAM/Knight
Brakes: AP/Lockheed

MINARDI M/85

Country of Origin: Italy
Date: 1985
Engine: Motori Moderni; V-6 Turbo; 720bhp at 11,300rpm; 2 × KKK turbochargers
Gears: Motori Moderni six-speed
Capacity: 1,498.9cc
Bore & Stroke: 80.0 × 49.7mm
Chassis: Suspension (front) double wishbone with

pull rods; suspension (rear) double pushrods. Suspension dampers Koni
Dimensions: wheelbase 260cm (102.6in). Track (front) 181cm (71.3in); track (rear) 166cm (63.39in)
Tyres: Pirelli
Steering: Minardi rack and pinion
Brakes: Brembo

ALFA ROMEO 185 T

Country of Origin: Italy
Date: 1985
Engine: Alfa Romeo 890T; V-8; 700bhp at 11,800rpm
Gears: Alfa five-speed.
Capacity: 1,496cc
Bore & Stroke: 74 × 43.5mm
Chassis: Suspension (front) by pull rods; suspension (rear) by push rods. Suspension dampers by Koni

Dimensions: wheelbase 278cm (109in). Track (front) 170cm (70.47in); track (rear) 166cm (65in)
Tyres: Goodyear
Steering: Alfa
Brakes: Brembo

OSELLA FAIF

Country of Origin: Italy
Date: 1985
Engine: Alfa Romeo; V-8 Turbo 1984; 540bhp a 12,300rpm
Gears: Alfa Romeo six-speed
Capacity: 2,995cc
Bore & Stroke: 78.5 × 51.5mm
Chassis: Suspension (front and rear) by push rods.

Suspension dampers by Koni
Dimensions: wheelbase 275cm (108.46in). Track (front) 173cm (68.11in); track (rear) 160cm (63in)
Tyres: Pirelli
Steering: Osella
Brakes: Brembo

ZAKSPEED 841

Country of Origin: West Germany
Date: 1985
Engine: Zakspeed; 4 cylinder; 700bhp at 11,500rpm; KKK turbocharger
Gears: Hewland/Zakspeed five/six-speed
Capacity: 1,495cc
Chassis: suspension (front) double wishbones

inboard springs; suspension (rear) pullrod. Suspension dampers Koni
Dimensions: wheelbase 282cm (111in). Track (front) 180cm (70.8in) track (rear) 160cm (63in)
Tyres: Goodyear
Steering: Zakspeed rack and pinion
Brakes: AP

Index

Abecassis, George 65
Andretti, Mario 165
Alfa Romeo P2 24
   Tipo B 42
   Tipo 158 'Alfetta' 74, 85
   Tipo 159 32, 85
   184T 230
Allard Sprint 78
Allard, Sydney 79
Alta Voiturette 64
Appleton Special 44
Arrows A7 206
Ascari, Alberto 79, 95
Aston Martin F1 DBR4-250 106
ATS Tipo 100F1 126
   D5-Cosworth 198
Austin Twin-cam 58
Austro-Daimler Sascha 22
Auto-Union C-type 60

Baghetti, Giancarlo 127
Bauer, Wilhelm 17
Behra, Jean 101
Bira, Prince 63
Birrell, Gerry 149
BMW Formula 2 144
Boillot, George 20
Bonnier, Joakim 101, 131
Bordeu (driver) 109

Borzacchini, Baconin 49
Brabham BT3 Formula 1 118
   BT20 128
   BT53 208
Brabham, Jack 119, 129, 136
Brise, Tony 175
BRM P25 100
   Type 56 Formula 1 120
Broadley, Eric 136
Brooks, Tony 92
Bugatti Type 35 26
   Type 59 46

Campari, Caraliere 49
Castelotti, Euginio 95
Cannstatt-Daimler Pheonix 16
Chevron B16 146
Clark, Jim 123, 139
Cobb, John 41
Connaught 'B' Type 92
Connell, Ian 57
Cooper-Climax T53 GP 110
Cooper-Maserati T81 130
Crosslé 16F Formula Ford 148

Daytona 24-Hour race 151
de Angelis, Elio 215
de Cesaris, Andrea 215

de Knyff, René 19
De Tomaso Type 38
  Formula 1 156
Dodson, Charles 59
Dornier (designer) 145
Dreyfus, René 69
Driscoll (driver) 59
Ducarouge, Gerard 215
Dusenberg Wonder-Bread
  Special 36

Eagle T2G Formula 1 132
Elva Formula Junior 112
Ensign F1-N176 180
ERA B-Type 62
  E-type 70
ERA-Delage Grand Prix
  32

Fane, A.F.P. 51
Fangio, Juan 77, 97
Farman, Henry 19
Ferguson P99 Formula 1
  116
Ferrari 1246C4 Turbo 210,
  312B, 158
  312P, 164
  312T-4, 190
  375, 84
  500, 88
Fittipaldi, Emerson 161, 183

Fittipaldi FD/04 182
Fontes, Louis 52
Ford GT40 Mk4 134
Foyt, A.J. 135
Francois, Jean 69
Frazer-Nash Single Seater
  50

Giunti, Ignazio 159
Gonzalez, Frolan 85
Goodacre (driver) 59
Gordini Formula 1 102
Goux, Jules 20
Gurney, Dan 119, 133, 135
Gurney Weslake 132

Hailwood, Mike 167
Hawthorn, Mike 89
Hill GH1 174
Hill, Graham 127, 139, 175
Honda RA 300 F1 136
HWM Formula 2 80

Ickx, Jacky 159, 165, 197, 203

Jaguar C-type 86
  D-type 90

Lancia 91
  D50 94
  LC2/83 202

237

Lauda, Nikki 217
Ligier JS23 212
Lola T70 Mk 3B 150
Lotus Type 25 122
   49 138
   63 152
   81-Cosworth 198
   JPS 95T Formula 1 214
   Type 72 Formula 1
   160, 177

Mansell, Nigel 215
March 821-Cosworth 200
Martini MK19 Formula 2 184
Martini, Renato 185
Maserati 250F 96
   8CM 48
   4CLT/48s 83
   Milan 82
Matra-Cosworth MS80
   Formula 1 154
   MS 120 Formula 1 162
May, Michael 109
Mazda 717C 204
McLaren, Bruce 139, 143
McLaren-Ford M7 142
   TAG MP4/2 209, 216
Mercedes Benz W125 66
   W196 98
Mexican Grand Prix
   118, 131, 137
MG K3 Magnette 38
   KI Magnette 39

K2 two-seater 39
Mille Miglia 39, 69
Minardi M/85 228
Monaco Grand Prix 63,
   77, 97, 109, 163, 173,
   221
Monte Carlo Rally 221, 225
Montlhéry 41
Monza 25, 49, 93, 163, 211
Moss, Stirling 97, 105, 117
Multi-Union II 72

Napier-Railton 40, 73
Neubauer, Alfred 22
Nicols, Frank 113
Nurburgring 109, 147,
   211
Nuvolari, Tazio 39, 43
   49

Osella FAIF 232
Oulton Park Gold Cup
   117, 153

Panhard 70hp 18
Parnell, Reg 120
Parnelli VPJ4 176
Parnelli Jones, Rufus 177
Pescara Grand Prix 61
Peterson, Ronnie 165
Peugeot 7.6-litre 20
Phillippe, Maurice 177
Piquet, Nelson 209

Porsche 917–30 170
  956C 196
  F1 Type 804 124
Prost, Alain 217

RAC Hill-Climb
  Championship 79
RAC Tourist Trophy 39
Railton, Reid 41
Railton-Mobil 70
RAM 03 226
Redman, Brian 147, 165
Renault RE50 218
  RSO1 Formula 1 188
Reventlow, Lance 115
Rindt, Jochen 163, 169, 171
Robinson, Tony 131
Rodriguez, Pedro 131
Rolt, Tony 32
Roots-type supercharger
  28, 32, 58, 66, 74
Rosberg, Kiki 225
Rosemeyer, Bernd 61
Rosier, Louis 77
Regazzoni, Gianclaudio
  159

Salvadori, Roy 107
Sayer, Malcolm 87
Scarab Formula 1 114
Scheckter, Jody 187
Schenken, Tim 165
Senna, Ayrton 221
Shadow DN5 178

Shelsley Walsh 51, 59
Siffert, Josef 119
Silverstone 87, 107
Solitude Grand Prix 109, 119
Southgate, Tommy 179
Spanish Grand Prix 160
Squire Single-Seater 52
Stanguellini Formula
  Junior 108
Stewart, Jackie 155, 173
STP-Turbocar 140
Straight, Whitney 37
Stutz 'Black Hawk' 30
Sunbeam Grand Prix 28
Surtees, John 130, 137, 167
Surtees TS10 Formula 2
  166

Talbot-Lago Type 26C 76
Tambay, Patrick 218
Targa Florio 22, 27
Taylor, Geoffrey 65
Tecno PA123 Formula 1
  168
Toleman TG183B 220
Tresillian, Stuart 101
Trossi, Count 37, 55
Trossi-Monaco 54
Tyrrell P34 186
  012 Grand Prix 222
  006/2 172

Vale Special 56
Vanwall 104
Villoresi, Emilo 95
Von Delius (Driver) 61
von Trips, Wolfgang 109

Weslake, Harry 133
Whitehead, Peter 71
Williams FW09-Honda 224
  FW06-Ford 192

Zakspeed 841 234